SAMPLE PREPARATION FOR BIOMEDICAL AND ENVIRONMENTAL ANALYSIS

CHROMATOGRAPHIC SOCIETY SYMPOSIUM SERIES

CHIRAL SEPARATIONS
Edited by D. Stevenson and I. D. Wilson

CHROMATOGRAPHY AND ISOLATION OF INSECT HORMONES AND PHEROMONES
Edited by A. R. McCaffery and I. D. Wilson

RECENT ADVANCES IN CHIRAL SEPARATIONS
Edited by D. Stevenson and I. D. Wilson

RECENT ADVANCES IN THIN-LAYER CHROMATOGRAPHY
Edited by F. A. A. Dallas, H. Read, R. J. Ruane, and I. D. Wilson

SAMPLE PREPARATION FOR BIOMEDICAL AND ENVIRONMENTAL ANALYSIS
Edited by D. Stevenson and I. D. Wilson

SAMPLE PREPARATION FOR BIOMEDICAL AND ENVIRONMENTAL ANALYSIS

Edited by

D. Stevenson

The Robens Institute of Industrial and Environmental Health and Safety
University of Surrey
Guildford, Surrey, United Kingdom

and

I. D. Wilson

Zeneca Pharmaceuticals
Macclesfield, Cheshire, United Kingdom

PLENUM PRESS • NEW YORK AND LONDON

Library of Congress Cataloging-in-Publication Data

Sample preparation for biomedical and environmental analysis / edited
by D. Stevenson and I.D. Wilson.
     p. cm. -- (Chromatographic Society symposium)
  "Proceedings of a Chromatographic Society International Symposium
on Sample Preparation for Biomedical and Environmental Analysis,
held July 23-25, 1991, at the University of Surrey, Guildford
Surrey, United Kingdom"--T.p. verso.
  Includes bibliographical references and index.
  ISBN 0-306-44663-4
  1. Chromatographic analysis--Congresses. 2. Sampling--Congresses.
I. Stevenson, D. (Derrick) II. Wilson, Ian D. III. Chromatographic
Society (Great Britain) IV. Chromatographic Society International
Symposium on Sample Preparation for Biomedical and Environmental
Analysis (1991 : University of Surrey). V. Series.
  [DNLM: 1. Chromatography--methods--congresses. 2. Chemistry,
Pharmaceutical--methods--congresses. QD 79.C4 S192 1993]
QP519.9.C47S25 1993
574.19'285'028--dc20
DNLM/DLC
for Library of Congress   93-46629
                                            CIP

Proceedings of a Chromatographic Society International Symposium on Sample Preparation for
Biomedical and Environmental Analysis, held July 23–25, 1991, at the University of Surrey,
Guildford, Surrey, United Kingdom

ISBN 0-306-44663-4

©1994 Plenum Press, New York
A Division of Plenum Publishing Corporation
233 Spring Street, New York, N.Y. 10013

Printed in the United States of America

PREFACE

This volume represents the proceedings of an international symposium on sample preparation, held at the University of Surrey, and jointly organised by the Chromatographic Society and the Robens Institute.

The Chromatographic Society is the only international organisation devoted to the promotion of, and the exchange of information on, all aspects of chromatography and related techniques.

With the introduction of gas chromatography in 1952, the Hydrocarbon Chemistry Panel of the Hydrocarbon Research Group of the Institute of Petroleum, recognising the potential of this new technique, set up a Committee under Dr S.F. Birch to organise a symposium on "Vapor Phase Chromatography" which was held in London in June 1956. Almost 400 delegates attended this meeting and success exceeded all expectation. It was immediately apparent that there was a need for an organised forum to afford discussion of development and application of the method and, by the end of the year, the Gas Chromatography Discussion Group had been formed under the Chairmanship of Dr A.T. James with D.H. Desty as Secretary. Membership of this Group was originally by invitation only, but in deference to popular demand, the Group was opened to all willing to pay the modest subscription of one guinea and in 1957 A.J.P. Martin, Nobel Laureate, was elected inaugural Chairman of the newly-expanded Discussion Group.

In 1958 a second Symposium was organised, this time in conjunction with the Dutch Chemical Society, and since that memorable meeting in Amsterdam the Group, now Society, has maintained close contact with kindred bodies in other countries, particularly France (Groupement pour l'Avancement des Methodes Spectroscopiques et Physico-chimiques d'Analyse) and Germany (Arbeitskreis Chromatographie der Gesellschaft Deutscher Chemiker) as well as interested parties in Eire, Italy, The Netherlands, Scandinavia, Spain and Switzerland. As a result chromatography symposia, in association with instrument exhibitions, have been held biennially in Amsterdam, Edinburgh, Hamburg, Brighton, Rome, Copenhagen, Dublin, Montreux, Barcelona, Birmingham, Baden-Baden, Cannes, London, Nurnburg, Paris, Vienna, Amsterdam and Aix-en-Provence.

In 1958 "Gas Chromatography Abstracts" was introduced in journal format under the Editorship of C.E.H. Knapman: first published by Butterworths, then by the Institute of Petroleum, it now appears as "Gas and Liquid Chromatography Abstracts" produced by Elsevier Applied Science Publishers and is of international status - abstracts, covering all aspects of chromatography, are collected by Members from over 200 sources and collated by the Editor, Mr E.R. Adlard assisted by Dr P.S. Sewell.

Links with the Institute of Petroleum were severed at the end of 1972 and the Group established a Secretariat at Trent Polytechnic in Nottingham, Professor Ralph Stock playing a prominent part in the establishment of the Group as an independent body. At the same time, in recognition of expanding horizons, the name of the organisation was changed to the Chromatography Discussion Group.

In 1978, the "Father" of Partition Chromatography, Professor A.J.P. Martin was both honoured and commemorated by the institution of the Martin Award which is designed as testimony of distinguished contribution to the advancement of chromatography. Recipients of the award include:

E.R. Adlard	Prof. G. Guiochon	Prof. E. Jellum
Prof. D.W. Armstrong	Prof. J.F.K. Huber	Prof. A. Liberti
Prof. U.A.Th. Brinkman	Dr C.E.R. Jones	Dr C.S.G. Phillips
Prof. E. Bayer	Prof. J. Jorgenson	Prof. W.H. Pirkle
Prof. Carel A.M.G. Cramers	Dr R.E. Kaiser	Dr G. Schomburg
Prof. D. Desty	Prof. B. Karger	Dr R.P.W. Scott
Prof. L.S. Ettre	C.E.H. Knapman	Prof. L.R. Snyder
Prof. D.E. Games	Prof. J.H. Knox	Prof. R. Stock
Prof. J.C. Giddings	Prof. E. Kovats	Dr G.A.P. Tuey

The Group celebrated its Silver Jubilee in 1982 with the 14th International Symposium held, appropriately, in London. To commemorate that event the Jubilee Medal was struck as means of recognising significant contributions by younger workers in the field. Recipients of the Jubilee Medal include Dr J. Berridge, Dr H. Colin, Dr K. Grob Jr., Dr J. Hermannson, Dr E.D. Morgan, Dr P.G. Simmonds, Dr P. Schoenmachers, Dr R. Tijssen, Dr K.D. Bartle and Dr H. Lingeman.

In 1984 the name was once again changed, this time to The Chromagraphic Society, which title was believed to be more in keeping with the role of a learned society having an international membership of some 1000 scientists drawn from more than 40 countries. At that time, the Executive Committee instituted Conference and Travel Bursaries in order to assist Members wishing to contribute to, or attend, major meetings throughout the world.

The Society is run by an Executive Committee elected by its members, in addition to the international symposia, seven or eight one-day meetings covering a wide range of subjects are organised annually. One of these meetings, the Spring Symposium, is coupled with the Society's Annual General Meeting when, in addition to electing the Society's Executive Committee, members have the opportunity to express their views on the Society's activities and offer suggestions for future policy.

Reports of the Society's activities, in addition to other items of interest to its members (including detailed summaries of all papers presented at its meetings), are given in the Chromatographic Society Bulletin which is produced quarterly under the editorship of Dr W.J. Lough.

At the time of writing three grades of membership are offered: Fellowship Membership and Student Membership. Members receive the Bulletin free of charge, benefit from concessionary Registration Fees for all Meetings and Training Courses and are eligible to apply for Travel and/or Conference Bursaries.

For further information and details of subscription rates please write to:

Mrs J.A. Challis/Mrs G. Caminow
The Chromatographic Society
Suite 4, Clarendon Chambers
32 Clarendon Street
NOTTINGHAM
NG1 5JD. UK.
Tel: (0602) 500596

THE ROBENS INSTITUTE OF INDUSTRIAL AND ENVIRONMENTAL HEALTH AND SAFETY

The Robens Institute is Europe's largest University-based health and safety organisation, with an international reputation for its contribution to resolving problems resulting from our technological society. Since its foundation in 1978 the aims of the Robens Institute have been to advance health and safety worldwide by contributing in three ways:

1. By providing direct practical help to governments, industry, commerce and others with a particular health and safety problem.

2. By conducting longer-term research into fundamental health and safety issues.

3. By acting as a focus for discussion and training on health and safety matters.

In the desire for a safer world - a matter of increasing public concern - three fronts can be identified. There are problems in the Environment, in the Workplace, and with man-made Products, and the Institute's three main divisions correspond to these areas. It offers independent expert assessment, backed by high quality laboratory and field investigations. The Institute's clearly established independence, charitable status and wide experience in successfully dealing with emotive and confidential issues makes it ideally suited to meet the challenges of the future.

The Institute's highly qualified staff include internationally recognised authorities in toxicology, occupational and environmental medicine, epidemiology, ergonomics, microbiology, human physiology, occupational hygiene, safety, analytical chemistry, industrial psychology, information science and statistics. Many of the staff are members of national or international committees concerned with environment, workplace, and product health and safety.

As a part of the University of Surrey, the Institute is also able to draw very readily on the University's extensive specialised resources, including its excellent library and computing facilities. The Institute has also established a network of consultants to supplement its own expertise. Thus it can provide a comprehensive service on health and safety issues. Current and future legislation, particularly arising from the European Community, will necessitate that expert advice on these health and safety issues assumes a high priority in all plans for the future. The Robens Institute provides a truly international service, supervising projects or collaborating with other organisations in all five continents.

Our effectiveness in being able to offer advice and practical solutions stems from extensive experience gained in fundamental and applied research in the field of health and safety. Among sponsors of this work are many government agencies, international organisations, large industrial concerns and charitable trusts.

The final key to continued progress in health and safety matters is Education and Training, and in recognition of this the Robens Institute offers very comprehensive programmes ranging from one-day seminars to modular MSc courses. These attract participants from all over the world.

The Robens Institute is sited on the campus of the University of Surrey at Guildford. The main Institute building contains twenty well-equipped laboratories and supporting accommodation, with laboratories and offices in other parts of the campus and on the adjacent University of Surrey Research Park. It is planned to unite all these into one building in the near future. In the centre of the south east of England but with motorway and rail links to the rest of the country, the Institute is ideally placed to provide a rapid and effective response to the needs of industry and government. For the international traveller the London airports of Heathrow and Gatwick are less than an hour's drive away.

CONTENTS

ON-LINE SAMPLE TREATMENT FOR COLUMN LIQUID CHROMATOGRAPHY

U.A.Th. Brinkman

"Free Universiteit"
Department of Analytical Chemistry
De Boelelaan 1083
1081 HV Amsterdam
The Netherlands

SUMMARY

A major part of modern analytical problem solving deals with the trace-level determination of (in)organic constituents in complex environmental and biological samples by means of high performance liquid chromatography (HPLC). In such cases, on-line sample pretreatment and post-column reaction detection are steps of crucial importance, because HPLC displays a separation efficiency and detection sensitivity that do not really match those obtained in capillary gas chromatography. In this paper on-line sample treatment is discussed.

In the field of precolumn techniques, trace enrichment and sample clean-up via the use of short columns packed with non-selective alkyl-bonded silica or polymer packings, or with selective metal-loaded or antibody-loaded material is discussed. Attention is given to the set-up of automated systems, the use of several precolumns in series with HPLC combined on-line with capillary GC, and also to the use of on-line dialysis and electrodialysis units and the introduction of highly efficient filter disks. Developments in the area of on-line precolumn derivatisation are mentioned, and the potential of combined pre- and post-column strategies emphasised.

In all cases, applications are given to demonstrate the practicality of the various procedures and their usefulness for routine monitoring of environmental and/or biomedical samples.

INTRODUCTION

In modern trace-level organic analysis, chromatographic techniques play a predominant role. They are used to create an efficient separation of the analytes of interest in complex environmental or biological samples, prior to the actual measurement, i.e. the detection step. Unfortunately, however, even the combined force of an efficient separation and a

Sample Preparation for Biomedical and Environmental Analysis,
Edited by D. Stevenson and I.D. Wilson, Plenum Press, New York, 1994

sophisticated mode of detection does not always create sufficient selectivity and/or sensitivity for the final goal to be reached: quantification and/or identification of sample constituents typically present at the low ppm to low ppt (10^{-5} - 10^{-11} g/g) level. In such instances, special attention has to be devoted to sample pretreatment (for trace enrichment and clean-up) and pre- or post-separation derivatisation conversion of the analytes (for improved detection selectivity and/or sensitivity).

In recent years, column liquid chromatography (LC), and especially reversed-phase LC with its aqueous organic eluents, has achieved widespread acceptance and popularity. No doubt, this is at least partly due to the increasing attention being given to polar compounds, irrespective of whether these are drugs, pesticides, industrial chemicals or their breakdown products. Selectivity and sensitivity enhancement in LC by means of (on-line) post-column reaction detection has been the subject of many reviews and books [1,2] and will not be considered here. Rather, attention will be devoted to sample treatment for LC, and emphasis will be given to the use of on-line techniques. These are becoming increasingly important in all those situations where (i) large series of samples have to be analysed routinely, making rapid analysis, (semi-) automation and unattended operation aspects of major concern, and (ii) sensitive trace-level determination requiring the analysis of total samples or sample extracts rather than aliquots, under conditions in which analyte losses - due to, e.g. evaporation or irreversible sorption to vessel walls, and contamination - caused by the solvents or reagents used, laboratory air and/or sample manipulation in general - must be rigorously minimised.

In a majority of such cases, on-line sample treatment for LC is carried out by means of a so-called precolumn technique. Most attention will therefore be devoted to this alternative and, especially, to the role played by the nature of the precolumn packing material. In recent years, techniques such as on-line dialysis and electrodialysis have also received attention and they will be discussed accordingly. On-line precolumn derivatisation is another topic of current interest, partly because pre- and post-column strategies aimed at enhancing detection sensitivity and/or selectivity can be nicely compared. Conventional column-switching LC-LC procedures - which, in essence, are rather straightforward heart-cutting techniques - will not be considered. However, some remarks will be made concerning on-line LC-GC (GC: capillary gas chromatography), a topic of much current interest in which LC is used for either preseparation or trace enrichment, while the highly efficient separation-cum-detection is performed in the GC part of the set-up.

ON-LINE PRECOLUMN/LC

In an on-line precolumn/LC procedure, four main steps can be discerned (cf. Fig. 1):
- loading of the sample (typically an aqueous sample or aqueous extract) which results in trace enrichment of the analytes of interest, i.e. in increased sensitivity;
- flushing of the precolumn to wash out potentially interfering sample constituents, which ensures improved selectivity;
- desorption of the analytes from the precolumn - a step which should be rapid and efficient in order to ensure that the starting zone on the top of the analytical column will be sufficiently small;
- reconditioning of the precolumn, which is preferred with expensive precolumns and/or when reconditioning is rapid, or exchange of the precolumn cartridge which is recommended for all other situations.

Currently, there is an increasing tendency to use relatively small precolumns (typically 2-10 mm length x 2-4.6 mm ID). Packed precolumns are commercially available from several manufacturers, but annual slurry- or dry-packing of a precolumn does not present

2

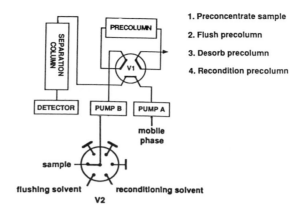

1. Preconcentrate sample
2. Flush precolumn
3. Desorb precolumn
4. Recondition precolumn

Figure 1.   Set-up of a precolumn sample pretreatment system for LC. V1, high-pressure switiching valve. V2, low-pressure solvent selection valve.

any real problems and, next to rapid exchange, allows easy screening of new packing materials. Recently, as an alternative to precolumns, so-called membrane extraction disks have been introduced [3,4] which contain a suitable stationary phase (C_{18}-bonded silica, ion-exchanger or copolymer; typically about 90 wt.%). The successful use of small 4.6 mm diameter disks for trace enrichment coupled on-line with LC has been reported [4].

In the majority of published procedures about on-line precolumn/LC, C_{18} or C_8 bonded silicas are used as the precolumn packing material. A nice illustration of the general usefulness of these materials is presented in Table 1 for a number of chlorophenols [5]. Obviously even compounds as hydrophilic as dichlorophenol will have breakthrough volumes of some 10 ml on a small precolumn, whilst for a really hydrophobic compound, such as pentachlorophenol, the breakthrough volume is over 300 ml. Although the latter figure is, of course, more impressive, one should realise that even a 10 ml sample loading represents a 100-fold improvement in sensitivity over a conventional 100 µl loop injection!

Although alkyl-bonded silicas are very effective for trace enrichment, their general drawback is that, together with the analytes of interest, many potentially interfering sample constituents will also be retained. Thus, selectivity will only be increased to a minor extent, if at all. In such instances, enhanced selectivity will have to be provided in the detection

Table 1.   Breakthrough volumes of selected chlorophenols on precolumns packed with alkyl-bonded silica or polymer materials.*

Packing Material	Breakthrough volume (ml) of chlorophenol				
	mono	*di*	*tri*	*tetra*	*penta*
LiChrosorb RP-18	0	12	58	180	320
Hypersil C-18	0	10	35	150	340
Polymer PRP$_1$	30	200			

*   2 mm x 4.6 mm ID precolumn; sample, water, pH 3.

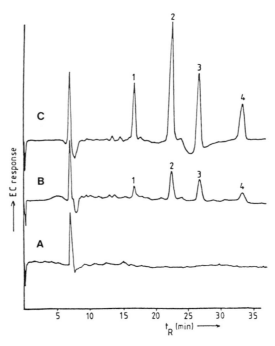

Figure 2. LC-electron capture detection chromatogram of 40 ml tap water after trace enrichment. (A) non-spiked, (B) spiked with 15-45 ppt and (C) spiked with 60-180 ppt of (1) 2,3,6-tri, (2) penta, (3) 2,3,5-tri and (4) 2,3,4-tri-chlorophenol. Normal-phase LC on silica with hexane - toluene - glacial acetic acid (79:20:1) as eluent.

step. One example is given in Figure 2, which shows the trace-level determination of several highly chlorinated phenols in tap water at the 10 to 100 ppt level using trace enrichment combined with normal-phase LC and electron-capture detection [6]. An alternative means to improve selectivity is to use post-column reaction detection. In one such study [7,8], N-methylcarbamates were preconcentrated on C_{18}-bonded silica, separated by conventional reversed-phase LC, the LC effluent then being led through a solid-phase reactor containing an anion-exchange resin which was kept at 100°C. The carbamates decomposed and methylamine was formed which reacted with orthophthalaldehyde in a second open-capillary-type reactor. The highly fluorescent reaction product was then monitored directly, because orthophthalaldehyde itself does not fluoresce. The determination of carbaryl after its trace enrichment from 20 ml of surface water demonstrates that both selectivity and sensitivity (detection limit, ca. 20 ppt) are excellent (see Figure 3).

If analytes cannot be satisfactorily preconcentrated on an alkyl-bonded silica - as is the case with, e.g. monochlorophenols and other hydrophilic organic compounds - one approach is to use a more hydrophobic packing material such as the styrene-divinylbenzene copolymers PLRP-S or PRP-1 (cf. Table 1). Generally speaking, this is a good approach which, to quote one example, is being explored [9] for the trace-level determination of polar pollutants in surface water. However, one should again realise - even more than with the bonded silicas - that, although sensitivity may be dramatically improved, selectivity is not. Besides, the analytical LC separation itself will normally require the use of a similar copolymer-type column which tends to have a lower efficiency than do alkyl-bonded silica columns.

Obviously, if precolumns have to be used in order to also enhance selectivity, analyte retention should be based on a principle providing better selectivity than hydrophobic

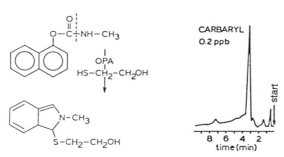

Figure 3.   Catalytic hydrolysis of carbaryl, its reaction with OPA, and the LC determination of 0.2 ppb of carbaryl in a 20-ml surface water sample using on-line trace enrichment and post-column reaction detection.

interaction - that is on, e.g. ion exchange, ligand exchange, chelate formation or antibody-antigen interaction. Two examples may serve to illustrate this approach.

Metal-loaded packing materials can easily be prepared by flushing an (inexpensive) thiol- or 8-hydroxyquinoline-containing polymethylmethacrylate polymer with an excess of an aqueous solution of a suitable metal salt, e.g. silver nitrate. Such Ag(l)-loaded precol-umns have been used to preconcentrate pyrimidine nucleobases such as uracil and thymine, and structurally related compounds such as the drugs 5-fluorouracil and AZT, and the pesti-cide bromacil [10]. The precolumn set-up used for the trace-level determination of AZT in plasma is shown in Figure 4. Initially, the AZT-containing sample was loaded directly on the metal-loaded precolumn (sample pH 5). Desorption with a very small (60 µl) volume of 0.1 M perchloric acid effected the rapid and quantitative desorption of the analyte from the precolumn and its transfer to a conventional reversed-phase LC system. Even after pro-longed use, the repeated injection of the plug of strong acid did not cause noticeable deterio-ration of the performance of the analytical column. However, the precolumn procedure did not remove all endogenous compounds having the -NH(=CO) structure responsible for reten-tion. For that reason, a PLRP-S copolymer precolumn was inserted (Figure 4; No.5) to effect additional clean-up. Under these conditions, AZT could be determined in plasma down to a concentration of 10^{-8} M using UV detection at 269 nm.

The use of immobilised antibodies for selective on-line sample treatment in LC has recently been reported for, e.g. the anabolic hormone β-19-nortestosterone and its main metabolite in calf urine and meat [11]. Other applications include the determination of β-trenbolone, chloramphenicol, clenbuterol and aflatoxin M_1 [12]. After sorption of the ana-lyte(s) of interest from, typically, a biological fluid or milk onto a properly pretreated immunoaffinity precolumn, on-line desorption is carried out either selectively, i.e. by using an essentially aqueous solution containing a so-called displacer (norgestrel in the case of nortestosterone), or non-selectively with a methanol-water or an acetonitrile-water mixture. In the latter instance, additional water has to be pumped in between the immunoaffinity precolumn outlet and the inlet of a second precolumn packed with an alkyl-bonded silica to allow refocusing of the analyte-containing zone on the latter precolumn. In most studies publihed so far, detection limits are well below the 1 ppb level. Some of the immunoaffinity precolumns have been used for well over 100 runs. One relevant application [13] viz. for aflatoxin M_1 in milk is shown in Figure 5.

ALTERNATIVE ON-LINE SAMPLE TREATMENT TECHNIQUES

In recent years, the use of dialysis as an on-line sample treatment technique for the removal of macromolecules prior to LC has received much interest. A dialysis module

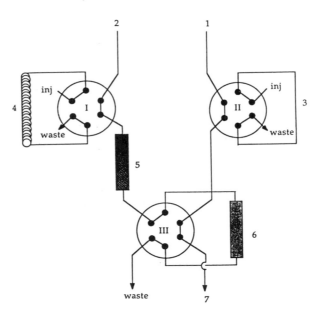

Figure 4. Scheme of the analytical system used for the determination of AZT in biological samples. (1) LC pump, (2) preconcentration pump, (3) injection loop (perchloric acid), (4) sample injection loop, (5) polymer-based clean-up precolumn, (6) Ag(I)-thiol precolumn, (7) to LC column.

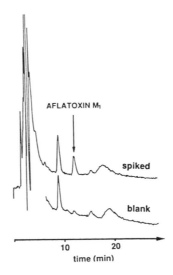

Figure 5. LC-fluorescence detection of a crude milk sample spiked with 50 ng/l of aflatoxin M_1, and of the corresponding blank. Hollow fiber dialysis/immunoaffinity preconcentration was performed with donor and acceptor volumes of each 25 ml.

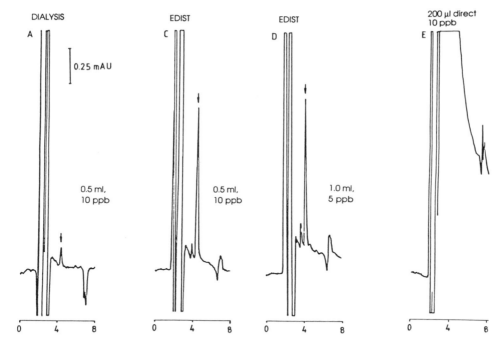

Figure 6.   LC-UV chromatograms of ground water samples spiked with 5-10 ppb of paraquat (see arrow). Pretreatment by means of dialysis and electrodialysis (EDIST) is compared with direct injection.

consists of two perspex blocks with the dialysis membrane (cut-off typically 10-15 kD) in between to separate the donor (sample) from the acceptor phase, It is important to realise that depending on the aim and the boundary conditions of the analysis (high recovery, rapid analysis, large or limited sample volume), different modes of operation should be selected: static/static, static/flowing, pulsed flow/flowing or continuous flow/flowing. Since, in practice, a flowing acceptor stream is almost invariably used, after dialysis the analytes have to be reconcentrated on a short (often C_8 or C_{18} type) precolumn prior to analysis by LC. Relevant examples include the determination of sulphonamides [14] and nitrofurans [15] in aqueous extracts of eggs, meat and milk, and of oxytetracycline [16] in salmon plasma and whole blood. Hollow-fibre dialysis has been shown to be highly successful in the determination of traces of aflatoxins in milk by immunoaffinity precolumn/LC [13].

Very recently, first results have been published [17] concerning electrodialytic sample treatment, where 5 to 10 volts are applied acros the dialysis cell (now containing electrode compartments and ion-exchange membranes next to the ordinary cut-off membrane). This promotes the transfer of charged analytes from the donor to the acceptor phase by means of electromigration. In such a case, a slowly flowing donor stream and a stagnant acceptor stream are used, and the dilution effect on the acceptor side, which invariably occurs with conventional dialysis, is absent. Rather, selective enrichment (up to 10 to 20-fold) of the charged analytes is achieved as has been shown for, e.g. the determination of several aromatic sulphonic acids in river Rhine water, and of paraquat and diquat in ground water (Figure 6) [18].

A survey of the literature shows that, whereas post-column reaction detection is invariably carried out on-line, pre-column labelling or conversion of analytes is usually done off-line. In recent years, however, on-line pre-column derivatisation has attracted some attention. Most studies deal with the low-level determination of primary and/or secondary amino acids in a variety of samples, using reagents such as 9-fluorenylmethyl chloroformate

7

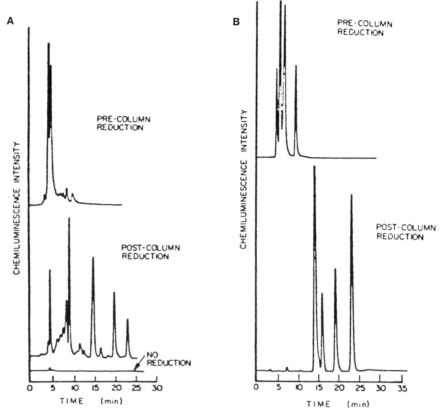

Figure 7. LC with chemiluminescence detection of (left) a carbon black extract and (right) a nitrated pyrene sample. The nitrated polynuclear aromatics are detected as their amino analogues after on-line pre- or post-column reduction with powdered zinc contained in a small reactor.

or othophthalaldehyde [19,20]. Modern autosamplers or autosampler-related devices are used to carry out the required reactions under such (geometrical) conditions that on-line coupling with the LC part of the system is possible.

When utilising orthophthalaldehyde as reagent, the addition of a mercapto reductant containing a chiral centre allows the separation of D-and L-amino acids - viz., as their diastereomers - on a conventional alkyl-bonded silica column [21]. Another interesting application [22] is the separation of nitrated polyaromatics - for example, nitrated pyrenes - as their amino analogues. Reduction is carried out on a precolumn filled with a mixture of small zinc particles and glass beads. The (chemiluminescence) detection limits are between 0.1 and 10 pg. It is interesting to add that the reduction can also be carried out successfully in the post-column mode (Figure 7). Phase-transfer catalysis has also been used for on-line pre-column analyte derivatisation. In one study, solutes such as ethinylestradiol and 4-monochlorophenol were determined in urine samples [23]. Dansylation in an aqueous-organic two-phase system was followed by on-line LC separation and fluorescence detection. In another paper [24], a micellar system was used for the rapid derivatisation of the drug valproic acid and free fatty acids in plasma using substituted coumarins and acridines as reagents. Under suitable conditions, the reaction is complete within about 10 mins, and on-line coupling to an LC system presents no serious problem.

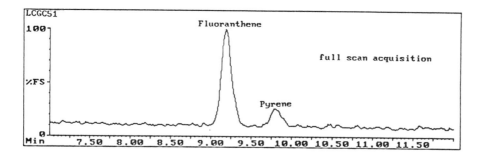

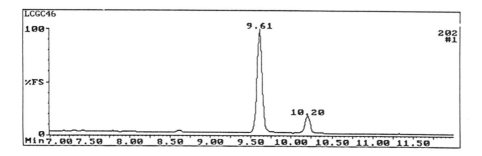

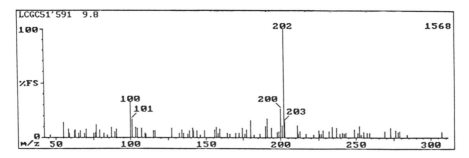

Figure 8. On-line LC-capillary GC with MS detection of polynuclear aromatics in an olive oil sample (low ppb level). Part of the final GC chromatogram is shown in the full scan acquisition and the selected ion monitoring (m/z 202) mode. The mass spectrum of pyrene is also shown.

ON-LINE LC-GC

Another successful approach to sample treatment is the on-line coupling of LC and capillary GC. Admittedly, in an LC-GC system, the role of LC is reduced to that of providing sample treatment while GC provides the real separation. However, the on-line coupling of LC with, in principle, the full range of universal as well as selective GC detectors, makes LC-GC too good a combination to be missed in an overview on pretreatment of (often aqueous) samples.

An exhaustive discussion of the various LC-GC interfaces currently available is inappropriate here (see, for example, refs. 25 and 26). Here, some interesting recent applications will be briefly discussed.

In one (heart-cutting) study [27], olive oil (diluted with hexane) was injected onto a normal-phase LC system; pentane containing 2% of methyl-tert.-butylether was used as eluent. The PAH-containing fraction was transferred on-line to a GC-MS system and analysed. A relevant result is shown in Figure 8. Detection limits for the individual PAHs in the SIM mode were as low as 1 pg. In another (trace enrichment) study (Brinkman *et al*, unpublished observations), 1 ml of a surface water sample spiked with test solutes such as nitrobenzene, m-cresol and acenaphthene was loaded onto a 5x2 mm ID precolumn filled with PLRP-S copolymer packing material. After on-line desorption, with 75 µl of ethyl acetate, to a GC-FID system, all spiked solutes could be detected at the 0.1 ppb level. A further example [28] deals with the determination of the herbicide fenpropimorph in cereals. After a conventional reversed-phase LC separation, the proper heart-cut is mixed on-line with an organic solvent. A segmented stream results and the analyte of interest is extracted into the organic phase which is separated from the aqueous phase in a non-membrane-type phase separator, collected in a loop and, then, introduced into a GC-NPD system via a retention gap.

CONCLUSIONS

Sample treatment is often the bottleneck in modern organic trace-level analysis. Most procedures still are off-line in nature, and they tend to be tedious, time-consuming and prone to error. Since there is an increasing demand for the routine monitoring of ever larger numbers of samples, the development of on-line sample treatment procedures is critically important: sensitivity increases, losses are prevented and automation becomes readily accessible.

Today, both disposable and reusable precolumns filled with a variety of polymers, alkyl-bonded silicas or selective stationary phases are routinely used for on-line precolumn/LC. Many biomedical and environmental applications have been published and the robustness of such systems is highly satisfactory. Several alternative solutions are increasingly attracting attention. Amongst these, dialysis - with its efficient removal of high-molecular-weight material - already takes a predominant place. The on-line coupling of (reversed-phase) LC and capillary GC still poses a number of problems. However, the separation and detection power of an LC-GC system is sufficiently rewarding to justify the research activities required to solve these problems.

REFERENCES

1. U.A.Th. Brinkman, R.W. Frei and H. Lingeman. Post-column reactors for sensitive and selective detection in high-performance liquid chromatography: Categorisation and applications. *J. Chromatogr.*, 492, 251-298 (1989).
2. Reaction detection in liquid chromatography. I.S. Krull (ed), M. Dekker, New York, (1986).
3. D.F. Hagen, C.G. Markell, G. Schmidt and D.B. Blevins. Membrane approach to solid-phase extractions. *Anal. Chimica Acta*, 236, 157-164 (1990).
4. E.R. Brouwer, H. Lingeman and U.A.Th. Brinkman. Use of membrane extraction disks for on-line trace enrichment of organic compounds from aqueous samples. *J. Chromatogr.*, 29, 415-418 (1990).
5. C.E. Werkhoven-Goewie, U.A.Th. Brinkman and R.W. Frei. Trace enrichment of polar compounds on chemically bonded and carbonaceous sorbents and application to chlorophenols. *Anal. Chem.*, 53, 2072-2080 (1981).
6. F.A. Maris, J. Stäb, G.J. de Jong and U.A.Th. Brinkman. On-line trace enrichment on a reversed-phase pre-column for normal-phase liquid chromatography with electron-capture detection. *J. Chromatogr.*, 445, 129-138 (1988).
7. L. Nondek, R.W. Frei and U.A.Th. Brinkman. Heterogeneous catalytic post-column reaction detectors for high-performance liquid chromatography application to N-methylcarbamates. *J. Chromatogr.*, 282, 141-150 (1983).
8. K.S. Low, U.A.Th. Brinkman and R.W. Frei. Liquid chromatographic residue analysis of carbaryl based on a post-column catalytic reactor principle and fluorigenic labelling. *Anal. Letts.*, 17, 915-931 (1984).

9. I. Liska, E.R. Brouwer, A.G.L. Ostheimer, H. Lingeman, U.A.Th. Brinkman, R.B. Geerdink and W.H. Mulder. Rapid screeing of a large group of polar pesticides in riverwater by on-line trace enrichment and column liquid chromatography. *Intern. J. Environ. Anal. Chem.*, 47, 267-291 (1992).

10. H. Irth, R. Tocklu, K. Welten, G.J. de Jong, U.A.Th. Brinkman and R.W. Frei. Trace-level determination of 3-azido-3- deoxythymidine in human plasma by preconcentration on a silver(I)-thiol stationary phase with on-line reversed-phase high-performance liquid chromatography. *J. Chromatogr.*, 491, 321-330 (1989).

11. A. Farjam, G.J. de Jong, R.W. Frei, U.A.Th. Brinkman, W. Haasnoot, A.R.M. Hamers, R. Schilt and F.A. Huf. Immunoaffinity pre-column for selective on-line sample pre-treatment in high-performance liquid chromatography determination of 19-nortestosterone. *J. Chromatogr.*, 452, 419-433 (1988).

12. A. Farjam. *Ph.D. Thesis*, Vrije Universiteit, Amsterdam, The Netherlands (1991).

13. A. Farjam. N.C. van de Merbel, A.A. Nieman, H. Lingeman and U.A.Th. Brinkman. Determination of aflatoxin M1 using a dialysis-based immunoaffinity sample pretreatment system coupled on-line to LC: re-usable immunoaffinity columns. *J. Chromatogr.*, 589, 141-149 (1992).

14. M.M.L. Aerts, W.M.J. Beek and U.A.Th. Brinkman. Monitoring of veterinary drug residues by a combination of continuous flow techniques and column-switching high-performance liquid chromatography. *J. Chromatogr.*, 435, 97-112 (1988).

15. M.M.L. Aerts, W.M.J. Beek and U.A.Th. Brinkman. On-line combination of dialysis and column-switching liquid chromatography as a fully automated sample preparation technique for biological samples. *J. Chromatogr.*, 500, 453-468 (1990).

16. T. Agasoster and K.E. Rasmussen. Fully automated high-performance liquid chromatographic analysis of whole blood and plasma samples using on-line dialysis as sample preparation. Determination of oxytetracycline in bovine and salmon whole blood and plasma. *J. Chromatogr.*, 570, 99-107 (1991).

17. A.J.J. Debets, W.Th. Kok, K.-P. Hupe and U.A.Th. Brinkman. Electrodialytic sample treatment coupled on-line with high-performance liquid chromatography. *Chromatographia*, 30, 361-366 (1990).

18. A.J.J. Debets, W.Th. Kok, K.-P. Hupe and U.A.Th. Brinkman. Electrodialytic sample treatment coupled on-line with high-performance liquid chromatography. *Chromatographia*, 30, 361-366 (1988).

19. R. Schuster. Determination of amino acids in biological, pharmaceutical, plant and food samples by automated precolumn derivatisation and high-performance liquid chromatography. *J. Chromatogr.*, 431, 271-284 (1988).

20. R.M. Kamp. High-sensitivity amino acid analysis using high performance liquid chromatography and precolumn derivatisation. *LC-GC Intern.*, 4, 40-46 (1991).

21. R.H. Buck and K. Krummen. Resolution of amino acid enantiomers by high-performance liquid chromatography using automated pre-column derivatisation with a chiral reagent. *J. Chromatogr.*, 315, 279-285 (1984).

22. K.W. Sigvardson and J.W. Birks. Detection of nitro-polycyclic aromatic hydrocarbons in liquid chromatography by zinc reduction and peroxyoxalate chemiluminescence. *J. Chromatogr.*, 316, 507-518 (1984).

23. C. de Ruiter, J.N.L. Tai Tin Tsoi, U.A.Th. Brinkman and R.W. Frei. On-line procedures for the phase-transfer-catalysed dansylation of phenolic steroids - Application to biological samples. *Chromatographia*, 26, 267-273 (1988).

24. F.A.L. van der Horst, M.H. Post, J.J.M. Holt Hui and U.A.Th. Brinkman. Automated high-performance liquid chromatographic determination of plasma free fatty acids using on-line derivatisation with 9-bromomethylacridine based on micellar phase transfer catalysis. *J. Chromatogr.*, 500, 443-452 (1990).

25. K. Grob. LC for sample preparation in coupled LC-GC: a review. *Chimia*, 45, 109-113 (1991).

26. J.J. Vreuls, G.J. de Jong and U.A.Th. Brinkman. On-line coupling of liquid chromatography, capillary gas chromatography and mass spectrometry for the determination and identification of polycyclic aromatic hydrocarbons in vegetable oils. *Chromatographia*, 31, 113-118 (1991).

27. P. van Zoonen, G.R. van der Hoff and E.A. Hogendoorn. Reversed-phase LC-GC with a nitrogen-phosphorus detector using a loop-type interface combined with a sandwich-type phase separator. *J. High Res. Chromatogr.*, 13, 483-488 (1990).

SAMPLE PREPARATION USING SUPERCRITICAL FLUID EXTRACTION

Roger M. Smith and Mark D. Burford[*]

Department of Chemistry
Loughborough University of Technology
Loughborough
Leicestershire, LE11 3TU UK

SUMMARY

Sample preparation is often proved the most time consuming part of a chromatographic assay because it usually involves an extraction procedure with organic solvents and the subsequent need to evaporate the solvent to concentrate the extract.

These processes can often be speeded up by using a supercritical fluid most commonly carbon dioxide as the extraction medium. Fluids, have lower viscosities than liquids combined with high diffusivity so that the extraction process can be shortened. Changes in the extraction pressure can provide selectivity of extraction by altering the density of the fluid. A major advantage is that by reducing the pressure within the system the supercritical fluid can be allowed to evaporate to give directly a concentrated extract.

The application of supercritical fluid extraction (SFE) to medicinal herbs is described with particular reference to the plant feverfew.

INTRODUCTION

In many analytical determinations the sample preparation stage is by far the most time consuming step. Recent years have seen the introduction of a number of labour and cost saving innovations, such as solid phase extraction (SPE) cartridges, robotics and intelligent autosamplers. However, the majority of methods are still restrained by the need to use organic or aqueous solvents either to separate the sample from the matrix or from a SPE cartridge. These separation steps are usually relatively slow because of the poor diffusion rates of liquids. The selectivity of the extraction is often limited by the available solvents and frequently after the extraction step, the solution must be concentrated by evaporation before the assay stage. These largely manual processes are often time consuming and hence costly. In

---

* Current address: Energy and Environmental Research Center, University of North Dakota, Grand Forks, North Dakota, USA

Sample Preparation for Biomedical and Environmental Analysis,
Edited by D. Stevenson and I.D. Wilson, Plenum Press, New York, 1994

13

addition the use of organic solvents incurs not only the cost of purchase but also the costs of the disposal of the waste solvent. In many organisations there are efforts to phase out chlorinated solvents, such as chloroform and dichloromethane, as their use is considered to be hazardous and environmentally damaging.

An alternative sample preparation which has been widely publicised in recent years is supercritical fluid extraction (SFE) [1-4]. However, SFE has not yet been widely adopted despite extensive and often optimistic claims by the manufacturers. As a method SFE has considerable potential in sample preparation but it is not suitable for all samples or matrices and some understanding of its advantages and limitations is important to avoid inappropriate applications. It does not offer an immediate solution to all sample preparation problems and developing a useful method requires just as much attention to detail as is needed with solvent or SPE cartridge methods. Method development in some cases can even be slower than with other methods as most laboratories have limited experience in this area.

Supercritical fluid extraction is not a new technique but was first introduced in the 1950s and 60. SFE has been in regular industrial use as a preparative extraction method for many years [5,6]. The most notable applications have been for the extraction of caffeine from coffee beans and flavour ingredients from hops. In contrast the recent interest over the last few years has been in the potential of SFE in analytical applications of a wide range of diverse matrices.

Instead of an organic solvent SFE employs a compound, mixture or element above its critical point as a supercritical fluid extraction solvent. Under these conditions the fluid is a single phase. Although the increased application of pressure will cause an increase in density the fluid will not condense to a two phase system. Compared to alternative solvent-based techniques SFE is claimed to give faster extractions, with improved efficiency and a cleaner work-up. These advantages come about primarily because supercritical fluids have physical properties between those of gases and liquids, so have higher diffusion rates than liquids and lower viscosities [1]. The most widely used supercritical fluid is carbon dioxide. However, although carbon dioxide is safe to use and is relatively cheap, it is very non-polar and, depending on the density used, its soublising power is between that of hexane and benzene. These properties can be altered by the addition of small proportions of modifiers, usually methanol or acetonitrile. Water, trifluoroacetic acid and more polar additives have also been used in experimental studies. Although a wide range of other fluids have been used for SFE, including freons [7] and nitrous oxide [2], none has gained acceptance as a routine method. One of the advantages of SFE is that most of compounds used as supercritical fluids are gases at ambient conditions so that solvent removal is simply achieved by reducing the pressure.

The extraction power of the fluid can be adjusted by using pressure and temperature to change the density of the supercritical fluid as more dense fluids have greater solvating power. Increasing the temperature, at constant pressure, reduces the density of the supercritical fluid but increases the volatility of the analyte so that low molecular weight analytes are more readily extracted.

In addition considerable interest has been shown in the role of SFE as a coupled extraction system in which it is linked to chromatographic separations, for example, gas chromatography (GC) [4,8-10], or supercritical fluid chromatography (SFC) [11], so that no separate collection or concentration steps are required. The ready removal of the solvent is particularly advantageous in coupled-SFE-capillary GC as the sample can be condensed in a retention gap ready for direct chromatography.

The potential analytical applications of SFE are widespread and examples have been reported from many areas, flavours, essential oils, oleoresins, natural colours, medicinal herbs, monomers from polymers, contaminants from catalysts, pesticides from soils and oil tars from shales. The extraction of polymer additives is very successful and detailed studies

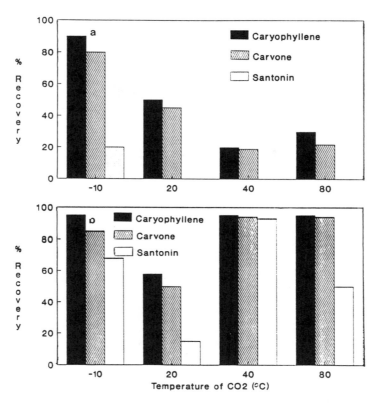

Figure 1.   Recovery yields of test compounds from spiked cellulose matrix. Conditions: carbon dioxide. SFE conditions: a, 250 kg cm^{-2} and b, 75kg cm^{-2}.

have demonstrated that the rate of extraction can be related to the diffusion rate through the bulk plastic [12]. Extraction takes place more rapidly from finely divided samples. With all these samples, care must be taken as it has been found that exhaustive extraction under constant conditions may not yield complete recovery of the analyte. Even if repeating the extraction does not yield an additional extract all the analyte may not have been obtained. Subsequent more strenuous conditions, such as the addition of modifier, can often yield additional material. It seems that for many analytes although part of the sample may be readily accessible, some of the analyte may be held is a less extractable form or location. Comparison of SFE with a second method is therefore an important stage in method development. In these cases a spiked sample may also give misleading results as the added analyte will probably be in an accessible form unlike the native analyte. Spiked samples, however, can play an useful role in testing recovery and sample trapping efficiency.

APPLICATION OF SFE TO MEDICINAL HERBS

An example of the variables that affect SFE methods can be demonstrated by work carried out at Loughborough on the extraction of medicinal herbs [13,14]. The aim of this study was to develop methods to confirm the identification of feverfew as a powdered dry plant material or in tablets and to distinguish it from powders of dried tansy and German chamomile which are possible adulterants. It was also desirable to be able to quantify the thermally labile sesquiterpene parthenolide which is reported to be the active ingredient.

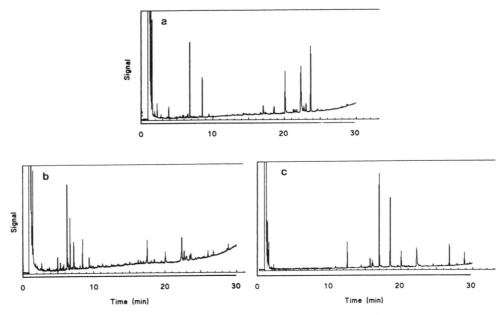

Figure 2.   GLC of SFE extracts from different herbs on 12 m x 0.33 mm BP1 fused silica column, temperature 60°C to 300°C at 8° min^{-1}. Extracts; a, feverfew; b, tansy and c, German chamomile.

Because initial studies suggested that the SFE extract was very different from a steam distil-
late and many of the volatile components were missing it was decided to work initially with
a model plant matrix and to examine the factors affecting extraction.

A limited group of terpenes and the sesquiterpene lactone santonin were spiked onto
α-cellulose [13]. This model matrix was then extracted at different pressures and temperatures with carbon dioxide and the extracts were examined by capillary GC. It was rapidly
determined that it was necessary to cool the SFE collection trap in liquid nitrogen to obtain
high recoveries of these volatile compounds.

Reasonable extractions were obtained using liquid (subcritical) carbon dioxide at
-10°C but yields decreased as the temperature was raised to 20°C, which could be attributed
to the reduction in density. If the temperature was raised to 40°C and pressures above the
critical pressure, so that supercritical carbon dioxide was being used as the solvent the extraction improved (Figure 1). At higher temperatures the yield was reduced because the
density of the carbon dioxide was reduced. The rate of extraction at 40°C and 250 kg cm^{-2}
was measured and it was found that extraction times of 20 to 30 min were needed for good
yields. Even then the more polar sesquiterpene santonin was poorly recovered. At 100 kg
cm^{-2} methanol had to be added to increase the yield above 40%.

The conditions of 40°C and 250 bar were then applied to dried feverfew and gave a
clean extract which could be compared with the very different extracts from tansy and
German chamomile (Figure 2) [14,15]. The extraction was much more rapid than from the
model system but without the addition of methanol only partial recovery of the parthenolide
was possible.

It was possible to selectively isolate a sample of parthenolide by placing a short silica
column after the extraction vessel (Figure 3A). On extraction with carbon dioxide the non-
polar terpenes and some of the parthenolide were transferred onto the silica trap column. The
terpenes were not retained by the silica and were eluted. The extraction vessel was then
switched out of the eluent stream (Figure 3B) and the silica column was eluted with carbon

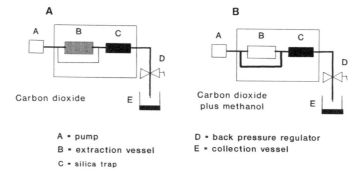

A = pump
B = extraction vessel
C = silica trap

D = back pressure regulator
E = collection vessel

Figure 3. SFE system fitted with silica trap to isolate parthenolide from feverfew. A, initial extraction conditions using carbon dioxide at 40°C and 200 kg cm^{-2}. B, extraction vessel switched out of the eluent flow and elution of parthenolide from the silica trap using carbon dioxide methanol (90:10).

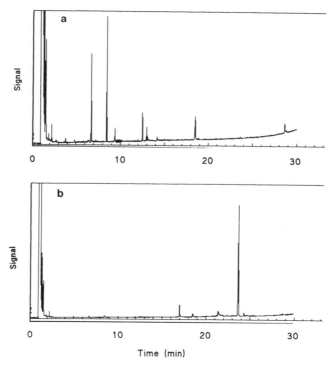

Time (min)

Figure 4. GLC of samples from selective extraction using silica column as in Figure 3. GLC conditions as Figure 2, parthenolide t_R 24 min.
Figure 4a, sample eluted with carbon dioxide at 250 bar and 40°C.
Figure 4b, sample eluted from silica trap with carbon dioxide + 10% methanol.

dioxide plus 10% methanol to give a 96% pure sample of parthenolide (Figure 4). Some parthenolide was left behind in the original plant sample but this loss was outweighed by the very ready isolation of the parthenolide compared to the traditional laborious extraction and chromatographic methods.

CONCLUSIONS

SFE offers a number of potential advantages as a sample preparation method in the analytical laboratory. These are primarily the speed of extraction and the ease of removal of the solvent. The conditions for extraction and solvent recovery are mild and can be made partially selective. In applying SFE the analyst needs to be able to compare the extraction with an alternative method and spend time validating the method as with any alternative technique. SFE is still a new technique and many laboratories still have limited experience of the method.

The main limitations of practical SFE are in the types of samples which can be readily analysed. Aqueous or liquid samples, such as blood and urine, are difficult to handle and so far little work has been carried out in these areas. Biological tissues pose a particular problem as the supercritical fluid is immiscible with water and cannot penetrate the bulk tissue material.

For many samples SFE appears highly suitable, particularly finely divided matrices, such as soil, sediments, ground plastics, animal feeds and dried plant material. It should find ready application in the pharmaceutical, agrochemical, polymer additive, and natural product areas where it can potentially save much tedious manual sample handling.

ACKNOWLEDGEMENTS

Thanks to the SERC for a studentship to MDB and to Ciba Corning for the loan of JASCO SFE/C equipment

REFERENCES

1. M.L. Lee and K.E. Markides (Eds.). "Analytical supercritical fluid chromatography and extraction", Chromatography Conferences, Provo, 1990.
2. S.B. Hawthorne. Analytical-scale supercritical fluid extraction. *Anal. Chem.*, 62, 663A (1990).
3. K. Sugiyama and M. Saito. Simple microscale supercritical fluid extraction system and its application to gas chromotography-mass-spectrometry of lemon peel oil. *J. Chromatogr.*, 442, 121 (1988).
4. M.R. Andersen, J.T. Swanson, N.L. Porter and B.E. Richter. Supercritical fluid extraction as a sample introduction method for chromotography. *J. Chromatogr. Sci.*, 27, 371 (1989).
5. A.B. Caragay. Supercritical fluids for extraction of flavors and fragrances from natural products. *Perfumer and Flavorist*, 6, 43 (1981).
6. D.F. Williams. Extraction with supercritical gases. *Chem. Eng. Sci.*, 36, 1769 (1981).
7. S.F.Y. Li, C.P. Ong, M.L. Lee and H.K. Lee. Supercritical fluid extraction and chromotography of steroids with Freon-22. *J. Chromatogr.*, 515, 515 (1990).
8. S.B. Hawthorne, M.S. Krieger and D.J. Miller. Analysis of flavor and fragrance compounds using supercritical fluid extraction coupled with gas chromotography. *Anal. Chem.*, 60, 472 (1988).
9. S.B. Hawthorne, D.J. Miller and M.S. Krieger. Coupled SFE-GC: a rapid and simple technique for extracting, identifying and quantitating organic analytes from solids and sorbent resins. *J. Chromatogr. Sci.*, 27, 347 (1989).
10. S.B. Hawthorne, D.J. Miller and M.S. Krieger. Rapid and quantitative extraction and analysis of trace organics using directly coupled SFE-GC. *J. High Resolut. Chromatogr.*, 12, 714 (1989).
11. M. Ashraf-Khorassani, M.L. Kumar, D.J. Koebler, and G.P. Williams. Evaluation of coupled supercritical fluid extraction-cryogenic collection-supercritical fluid chromatograph (SFE-CC-SFC) for quantitative and qualitative analysis. *J. Chromatogr. Sci.*, 28, 599 (1990).
12. K.D. Bartle, A.A. Clifford, S.B. Hawthorne, J.J. Langenfeld, D.J. Miller and R. Robinson. A model for dynamic extraction using a supercritical fluid. *J. Supercritical Fluids*, 3, 143 (1990).

13. R.M. Smith and M.B. Burford. Optimisation of supercritical fluid extraction of volatile constituents from a model plant matrix. *J. Chromatogr. Sci.*, (1992). In press.
14. R.M. Smith and M.B. Burford. Supercritical fluid extraction and gas chromatographic determination of the sesquiterpene lactone parthenolide in the medicinal herb feverfew (*Tanacetum partheium*), *J. Chromatogr. Sci.*, submitted for publication.
15. R.M. Smith and M.B. Burford. GLC of supercritical fluid extracts of essential oils from the medicinal herbs, feverfew, tansy and German chamomile, *J. Chromatogr. Sci.*, submitted for publication.

(POST-COLUMN) REACTION-DETECTION: AN ALTERNATIVE TO IMPROVE SENSITIVITY AND SELECTIVITY IN COLUMN LIQUID CHROMATOGRAPHIC ANALYSIS

H. Lingeman and U.A.Th. Brinkman

Department of Analytical Chemistry
Free University
De Boelelaan 1083
1081 HV Amsterdam
The Netherlands

SUMMARY

In sample preparation, in principle, two routes can be followed to improve selectivity and sensitivity. The most frequently applied is the rigorous removal of all interfering substances. This means, however, that in many cases time-consuming and laborious procedures must be used. The other route is the use of a specific detection technique for the analyte in the crude sample. Although this approach is not always possible chemical manipulations can be useful as a means of achieving this aim. Among these techniques, post-column reaction-detection systems are frequently used in combination with column liquid chromatographic (CLC) separations.

A wide variety of post-column reaction-detection techniques can be used. Examples are presented with respect to the use of on-line (electrochemical) generation of bromine for the detection of thiols, the hydrolytic reaction of barbiturates, the photolysis of dansylated phenols, the copper-containing reactor for the determination of thiram, and the immobilised enzyme reactor in combination with chemiluminescence detection.

INTRODUCTION

Although sample preparation is only a part of the total analytical procedure, about 61% of all the time spend on the analytical procedure is related to sample preparation. Moreover, 30% of all the errors generated during the total analytical procedure are caused by sample preparation [1]. With respect to sample preparation there are a few important aspects that should be kept in mind. First of all sample preparation should be considered as an integral part of the total analytical procedure, and secondly the lower the analyte concentration, the longer it takes to develop an assay.

Furthermore, it will be obvious that a sample-preparation procedure can have a number of objectives. The most important ones are [2,3]:

Sample Preparation for Biomedical and Environmental Analysis,
Edited by D. Stevenson and I.D. Wilson, Plenum Press, New York, 1994

- Removal of interferences;
- Improvement of instrumental resolution;
- Removal of material affecting the analysis;
- Solubilisation of analytes prior to instrumental analysis;
- Concentration of analytes within detection limits of the instrument;
- Dilution of sample to reduce the solvent strength, viscosity, or solvent incompatibility;
- Removal of material that blocks the frits, column, or injector;
- Stabilisation of analytes to avoid degradation.

These objectives are constrained by [2,3]:-

- The physico-chemical properties of the analytes;
- The analytical techniques available;
- The human expertise available;
- The time available;
- The matrices available.

This means that for successful method development [2,3]:-

- A clear objective of the method should be present;
- The analyst should understand the chemistry of the analyte and any associated reactions;
- The technical expertise of the staff should be known;
- The available instrumentation should be taken into consideration.

A significant number of sample preparation procedures are available and normally one or more methods are chosen of one of the first three groups of Table 1. However, this paper will be focused on the methods described in group 4 of this Table.

This is because in sample preparation, in principle, two approaches can be used [4]. The first one, and the most frequently applied one, is the complete removal of all interfering compounds followed by a rather non-selective separation-reaction-detection system. However, in a number of cases this means the incorporation of either time-consuming and/or laborious sample preparation steps. In the second approach the sample preparation step is totally or partly omitted and chemical manipulation reactions (e.g., pH changes, oxidation, reduction, hydrolysis, photolysis), large volume injections, or a specific pre- or post-column - separation-reaction-detection system is used directly on the crude sample to improve selectivity and/or sensitivity. Usually a combination of these two approaches are used.

Here methods of sensitivity improvement, including simple chemical manipulation reactions and post-column reaction-detection systems to improve selectivity and sensitivity without the need for additional pumps will be discussed.

IMPROVEMENT OF SENSITIVITY

Instrumental Factors

One of the simplest ways of increasing the sensitivity of a method is to inject large samples into the chromatographic system. Data on the injection of large volumes are described in the literature for biological samples [5]. This technique for concentration on top of the analytical column seems to be more effective in reversed-phase (RP) column liquid

Table 1.   Categorisation of sample preparation units.

Group 1	**Initial sample preparation techniques**	
	Dialysis Hydrolysis Precipitation Saponification Sonication (Ultra)Filtration	
Group 2	**Procedures for liquid handling**	
	Aspiration Centrifugation Dilution Evaporation Filtering Freezing Lyophilisation Mixing Pipetting Salting-out Separation	
Group 3	**Selective sample preparation techniques**	
	Liquid-liquid extraction Liquid-solid isolation Column liquid chromatography Immunoaffinity isolation Micellar chromatography Supercritical fluid extraction Electrophoresis	
Group 4	**Enhancement of selectivity and sensitivity**	
	Pre-column reaction/detection Post-column reaction/detection	Enzyme reactors Solid-phase reactors Ion-pair reactors Photochemical reactors Segmented-flow reactors Packed-bed reactors
	Selective detection modes	Diode-array detection Electrochemical detection Fluorescence detection Chemiluminescence detection

This Table is reprinted, with permission, from: R.D. McDowall, *J. Chromatogr.* 492:3 (1989).

chromatogaphy (CLC) compared with normal-phase (NP) CLC [6]. Examples include the determination of N-methylcarbamates in water, and the determination of metamitron in surface water by direct injection of a 2 ml sample (on to a 250 mm x 4.0 mm I.D. C_{18} column packed with 5 μm particles) [6]. Compared with a 20 μl injection no significant band broadening was observed, and the total analysis time was only 10 min. The detection limit, in surface water, was 0.15 μg/l using absorbance detection at 310 nm. Over 40 injections of 2 ml could be applied without significant loss in chromatographic performance.

Another simple way of improving the sensitivity is optimisation of the column dimensions [7]. Because, in general, concentration sensitive detectors are used, dilution (C_{max}/C_o) of the sample in the detector should ideally be kept as small as possible. This dilution can given by the following Equation (Eq.1):-

$$\frac{C_{max}}{C_o} = \frac{4 V_o \sqrt{N}}{\Pi \, dc^2 \, e \, L \, (1 + k') \, (\sqrt{2\Pi})} \tag{1}$$

in which V_o is the injection volume, N the efficiency (plate height), dc the internal diameter, L the length of the column, e the porosity of the sorbent, and k' the capacity factor of the analyte. Using this Equation it can be seen that when columns of 10-cm length and an internal diameter of 3 mm packed with 3 - 5 μm particles are used the dilution is significantly less than when using columns of 250 x 4.6 mm I.D. packed with particles of 5 or 10 μm are used, and so an improved sensitivity is obtained.

CHEMICAL MANIPULATION REACTIONS

The selectivity and/or the sensitivity of an analytical procedure can be enhanced by using a number of simple chemical manipulation procedures. In all of the following examples fluorescence detection of the reaction products is used. For example, hydroxy substituted carboxylic acids in urine can be selectively detected, after ion-exchange separation using post-column oxidation [8], while tryptophan can be detected after pre-column oxidation [9]. In order to determine morphine in biological fluids, both pre- and post-column oxidation techniques have been described [10,11]. Post-column reduction procedures have been described for the determination of vitamin K_1 in plasma samples. Using electrochemical reduction, the detectability is in the order of 0.1 ng/ml with a day-to-day precision of about 3.2% (Figure 1) [12]. For the detection of indomethacin in biological samples a post-column hydrolysis procedure [13] or a pre-column deacylation technique can used [14]. Post-column irradiation with ultraviolet light (cf. below) is a useful technique for the determination of phenolic cannabinoid derivatives with fluorescence detection after a NP separation [15]. Warfarin [16] and purines [17] can only be detected sensitively after post-column acid-base manipulation. Photoreduction is a nice technique to improve both sensitivity and selectivity for the determination of carbohydrates [18] and ethers [19] in biological fluids. Another way of improving detection selectivity is described for amoxicillin and ampicillin. These compounds are degraded before they are subjected RP-CLC and fluorescence detection [20]. A final example of a simple chemical manipulation reaction to improve detectability is the post-column heating of (nor)epinepfrine samples after ion-exchange separation [21]. Heating of samples can also be performed in the presence of an ethanol-sulphuric acid mixture or concentrated hydrochloric acid, as described for the determination of cortisol [22] and digitalis glycosides [23], respectively.

Figure 1. Chromatogram of a serum extract after intravenous administration of vitamin K1. Chromatographic conditions: C_{18} stationary phase; mobile phase, 6.25% ethyl acetate in methanol containing 0.03 M sodium perchlorate; flow rate, 0.67 ml/min; reduction potential electrodes, -700 mV; excitation and emission wavelength, 320 nm, 430 nm, respectively. 1, vitamin K1; 2, 2',3'-dihydrovitamin K1. (Reprinted with permission from reference 12).

ON-LINE REACTION-DETECTION SYSTEMS

The use of a specific separation-reaction-detection system means the incorporation of a chemical manipulation reaction into the analytical scheme. Both post-column and precolumn reaction-detection techniques can be used and both have their own advantages and limitations. The most important advantage of pre-column techniques is that there are no restrictions in the reaction kinetics, while side-product formation is one of its major problems. On-line and semi on-line pre-column reaction-detection is gradually becoming a worthwhile alternative to post-column techniques. So far, however, attention has been essentially restricted to the determination of amino acids in various matrices, but the potential of the approach is, of course, much wider. This, especially nowadays, since fully automated sample processors are available which can perform this type of derivatisation. An extended overview on the use of this type of reactions can be found in ref 24.

The post-column reaction-detection principle however, has advantages over precolumn techniques with respect to artefact formation and the possibility of simultaneous use of different detection principles. The major disadvantage of post-column methods lies in the restrictions in the choice of the chromatographic eluent, because this can strongly influence the reaction medium. Other technical problems include the difficulty of using gradient elution, the need to avoid complex or kinetically awkward reactions, the need for additional pumps, and the requirement that the reagent should not be detectable under the same conditions as the derivative. Many of these difficulties, however, can be reduced or eliminated by choosing the proper reagent.

The most important parameter influencing the background in a post-column reaction-detection system is the additional noise generated by the reagent pumps. For example, in a system for the derivatisation of the primary amine function of peptides with fluorescamine [24], followed by fluorescence detection, it is important to use completely pulseless pumps. This is because the background fluorescence of the reaction mixture tends to amplify even the smallest pulses of even a well-damped piston pump. Solutions for this problem are the use of syringe pumps and pneumatic pumps for delivery of the reagent(s), or even better, the use of the so-called pumpless systems.

Post-column Reaction-Detection Systems

In principle post-column systems are rather simple. They consist of an isocratic or gradient CLC system, a mixing T-piece, an additional reagent pump, a reactor, and some capillaries. Such a simple set-up is already sufficient for quite a number of applications. However, the material that is used for post-column reaction-detection system is often rather critical. The T-pieces are normally made of stainless steel and are commercially available, reaction capillaries can be made of glass, stainless steel or PTFE, which can be easily coiled or knitted, and the necessary phase-separators are also commercially available nowadays [24].

The most important part of a post-column reaction-detection system is the reactor. Nowadays only a limited number of reactors are widely used. These are the open-tubular, packed-bed, and segmented-stream reactors. The latest development in this area is the hollow-fibre membrane reactor. While the open-tubular reactor can only be used for fast reactions with residence times of less than 30 s, the packed-bed and segmented-stream reactors can be used for slower reactions with residence times up to 15 min. The packed-bed reactors can be used in different forms: a capillary packed with inert particles or a capillary packed with active particles to perform heterogeneous reactions. In packed-bed reactors, for example, the particle size of the particles and the type of sorbent determine the maximum residence time.

The most important aspect of post-column reactors is the additional band broadening, which can result in a loss of resolution. As can be seen in Figure 2 coiling and knitting of the capillaries significantly decreases the extra-column band broadening because an additional convective flow is introduced. However, nowadays band broadening in mixing-T-pieces is no longer a limiting factor. For example, mixing units are developed for the reaction of primary amines with o-phthaldialdehyde (OPA). They consists of two metal blocks, one of which has a groove containing a PTFE insert. The volume of this groove is 60 nl. In the other part there are three capillaries with an I.D. of 0.12 mm. The two capillaries on the outside are used as inlet for the two solvents into the mixing chamber and the capillary in centre is the outlet of the system.

Another important parameter is the influence of the flow rate on the extra-column band broadening. Because of wettability problems, the material of the capillary can be of major importance. For example, using a liquid segmented-stream system there is hardly any band broadening using a glass capillary, while the band broadening is strongly flow dependent using a PTFE capillary.

Phase separators are also a critical part of post-column reaction-detection systems, and much attention has been devoted to their proper construction. The major principle used for phase separation is wetting. Early designs are based on the insertion of PTFE tubing in a glass T-piece device. Efficient separation of an aqueous and organic phase is created by wetting of the glass and an hydrophobic (PTFE) surface, respectively. Gravity plays only a minor role in this type of phase separator.

More recently sandwich-type phase separators have been developed. The principle is again based on wetting. In Figure 3 the scheme of such a phase separator is shown. There is an internal (groove) volume of only ca. 40 μl between the two blocks of stainless steel. Phase separation occurs in the groove(s) machined in the upper stainless-steel block and the PTFE disc. Further miniaturisation to a groove volume of about 8 μl has already been achieved.

Using the post-column reaction-detection system (partly) as a sample treatment step, a selective detection mode should be used. Therefore, only fluorescence, amperometric, and

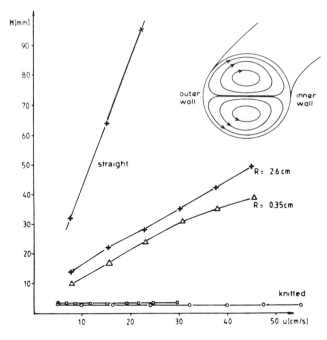

Figure 2. Peak dispersion in ideal, tightly coiled and knitted open tubes. Insert: secondary flow pattern in the cross-section of a coiled tube.

chemiluminescence detection devices will be described. As explained before the emphasis will be on the so-called pumpless systems. A limited number of illustrative examples will be discussed.

Electrochemical Reactions

Electrochemical reagent addition is one example of a technique allowing the introduction of an extremely pure reagent without the need of using additional pumps. An elegant approach is the on-line post-column generation of bromine or iodine. Subsequently, the bromine or iodine reacts with the analytes and the excess of halogen is detected amperometrically [25]. A specially constructed electrochemical cell should be used to avoid additional band broadening (Figure 4).

One recent application of this device is the determination of oxidised, reduced and protein-bound glutathione in eye lenses [26]. Homogenised lens samples were deproteinised with a mixture of perchloric acid and acetonitrile, and the protein-bound glutathione reduced with 1,4-dithiothreitol. Separation of the different forms of glutathione and dithiothreitol was performed by ion-pair RP-CLC using sodium octylsulphate as the ion-pairing agent. The compounds were detected amperometrically (at 300 mV) using on-line generated bromine, which oxidizes thiols and disulphides. The bromine was generated using a KOBRA cell [25,26], followed by reaction in a PTFE reaction coil (1700 mm x 0.5 mm) and detection in an amperometric detector. The minimum detectable amounts were about 0.5 pmol, and the repeatability was ca. 1.5%. The detection limits were 80 and 48 nmol/g wet lens for reduced and oxidised glutathione, respectively.

27

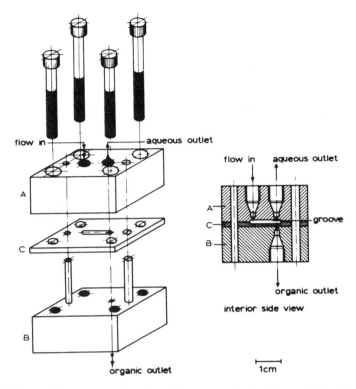

Figure 3.   Sandwich-type phase separator. Parts A and B constructed of stainless steel. Part C constructed of PTFE.

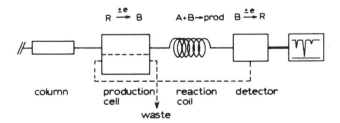

Figure 4.   Scheme of the on-line generation of bromine/iodine amperometric detection system. A, analyte; B, reagent (bromine/iodine); R, precursor of the reagent (bromide/iodide).

Hydrolytic Reactions

Hydrolytic reactions can be used in a simple form as described above or by using more advanced technologies such as the use of parallel-column systems [27]. A pronounced disadvantage of some non-catalytic solid-phase reactors is that they are rather rapidly depleted and thus can be used only for a short period of time before reloading becomes necessary. Another drawback is that often the reactor substrate is not available as regular and fine particles, which can result in severe band broadening. In certain cases these problems can be solved by employing a parallel reactor column approach, as has been demonstrated for the determination of barbiturates in biological fluids [28].

The analytical system comprised an anion-exchange column that was inserted parallel to the injection valve and the analytical column. One part of the acetate-containing mobile phase flowed through the injection valve and the analytical column to achieve separation, the other part flowed through the anion-exchange column where the acetate ions caused the release of hydroxide ions from this column. Finally the alkaline stream from the anion-exchange column was recombined with the analytical column effluent. This resulted in an alkaline medium which is favourable for the 254-nm absorbance detection of barbiturates. Using conventional CLC columns (100 mm x 3.0 mm I.D.) the detection limits of all the barbiturates tested were below 1 ng with a repeatability of 6.0%. Regeneration of the anion-exchange column (250 mm x 4.6 mm I.D.) was only needed after approximately 17 h. During these 17 h, 140 ml of alkaline solution was delivered and the pH remained essentially constant until it suddenly droped. Using exactly the same anion-exchange column in combination with a narrow-bore analytical column (100 mm x 1.0 mm I.D.), the detection limits were below 1.5 ng with a repeatability of 7.9%.

Photochemical Reactions

Photochemical reactions can be applied for various purposes, for example, to increase the detectability in absorbance, amperometric, fluorometric, or photoconductivity detection. However, the photochemical reactor is mainly used to convert weakly or non-fluorescent compounds into highly fluorescent products [29]. In a photochemical reactor the analyte is irradiated with a high- or medium-power UV light source (e.g., mercury, xenon-mercury). In addition to the lamp, a photochemical reactor consists of a quartz or, preferably, a PTFE reaction coil and a temperature-controlled reactor. Reaction times typically are in the order of 5 to 60 s.

The photochemical reactor is a typical pumpless reactor, as photons are in this case the only "reagent". Photochemical reactors are mainly used in the post-column mode because (i) the photochemical conversions are almost never quantitative and (ii) normally a number of products are formed. When using a photochemical reactor the main variable is the residence time of the analyte in the (open-tubular) reactor. The composition of the eluent, the wavelength of UV irradiation and the intensity of the light source play a less prominent role.

An example using the photochemical reactor is described by De Ruiter *et al* [30] for the enhanced fluorescence detection of dansyl derivatives of phenolic compounds in surface water. The photochemical decomposition by UV irradiation of dansyl derivatives, produced by an off-line pre-column derivatisation procedure, of phenolic compounds in methanol-water mixtures leads to the formation of the highly fluorescent dansyl-OH and dansyl-OCH$_3$ compounds. Using chlorophenols as model compounds, it was demonstrated that inductive effects, caused by the substituents, played a major role in the increase in the fluorescence signal (up to 8,000 fold) that was obtained after post-column UV irradiation of the dansyl derivatives. The optimum irradiation time for the dansyl derivatives was about 5.5 s. All

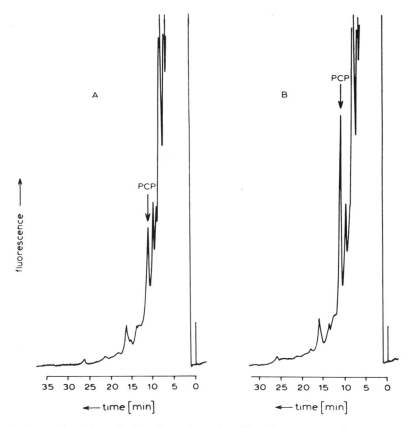

Figure 5.   Ultra trace level determination of pentachlorophenol in surface water. RP-CLC separation after on-line enrichment of 500 ml of sample. A, non-spiked surface water; B, surface water spiked with 100 pg/ml of pentachlorophenol.

monosubstituted phenolic dansyl derivatives had a comparable detection limit of approximately 200 pg. This means that the same calibration curve could be used for all such compounds. The calibration curve of the dansylated pentachlorophenol, using photolysis, was linear over at least three orders of magnitude with a correlation coefficient of 0.9999. The repeatability of the system for a surface water sample, spiked with 1 ng/ml of pentachlorophenol was 2.4%. Figure 5 shows a chromatogram of the ultra trace level determination of pentachlorophenol after on-line enrichment of 500 ml of surface water. The peak indicated in the chromatogram of the non-spiked sample corresponds to about 60 pg/ml of pentachlorophenol.

Ligand-Exchange and Complexation Reactions

Although ligand-exchange reactions are relatively slow, they can be applied in post-column reactors [31]. One application is the post-column complexation of thiram and disulfiram, for their selective RP-CLC determination using a metallic copper solid-state reactor [32]. The post-column complexation was performed on a solid-state reactor packed with finely divided metallic copper to form a coloured complex, copper(II)-N,N-dimethyldithiocarbamate, with an absorption maximum at 435 nm. During this reaction metallic copper

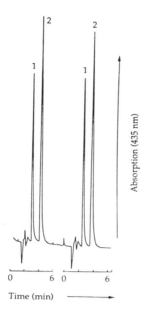

Figure 6. Chromatograms of the duplicate injections of surface water spiked with 10 µg/ml of thiram (1) and 20 µg/ml of copper(II) dimethyldithiocarbamate (2) using a post-column copper reactor and UV detection at 435 nm.

was oxidised to copper(II) and the disulphide reduced to a thiol. The method was combined with an on-line enrichment and clean-up step on a precolumn to permit the sub-ng/ml determination of the analytes in surface water samples.

The solid-state reactor (2.0 or 4.0 mm in length) is filled with metallic copper prepared by the reduction of copper(I) chloride. The resulting copper consists of fine particles which are pressed with a micro-spatula as densely as possible into the column. The reduction of disulphides with metallic copper, prepared in this way, is almost instantaneous. Using the 2 mm precolumn and 8 mg of reduced copper, flow rates up to 0.4 ml/min could be used, and over 200 injections applied, with a band broadening of less than 10%. In Fig. 6 chromatograms of the duplicate injection of surface water spiked with thiram and copper(II) dimethyldithiocarbamate, one of the main degradation products of thiram in the presence of copper(II) ions, are shown. Using a post-column reactor, these two analytes can be determined. After on-line enrichment of 40 ml of surface water, the detection limit for thiram was 5 ng/ml.

The same procedure can be used to determine disulfiram in urine samples. With the use of the copper-containing post-column reactor no complicated sample pretreatment was required. Filtered urine, containing disulfiram, was stabilised with an EDTA/citrate mixture. Then 1 ml of this sample was concentrated on a 4.0 mm long precolumn packed with C-18 bonded silica and then eluted to the CLC system with an acetonitrile-acetate buffer. With non-specific UV detection at 254 nm only one large peak could be seen in the chromatogram and the disulfirum could not be quantified. In contrast to this the use of the copper-containing post-column reactor and detection at 435 nm allowed sufficient separation of the disulfiram peak for accurate quantification.

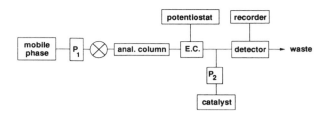

Figure 7.   Column liquid chromatographic - chemiluminescence system (CLC-CL) with on-line electrochemi
cal reagent generation. E.C., electrochemical flow cell (-600 mV); P1, mobile phase pump; P2,
microperoxidase pump. In the CLC-CL system with reagent addition, the E.C. is replaced by a third
pump, P3, for the addition of hydrogen peroxide in water.

Chemiluminescence Reactions

Chemiluminescence detection is one of the most sensitive detection modes in CLC. This is mainly because excitation of the analytes is performed by means of a chemical reaction, and as a result of a strongly decreased background, the signal-to-noise ratio is significantly improved. Various chemiluminescence reactions can be applied for detection purposes in CLC, but the most frequently used ones are the peroxyoxalate and luminol systems [33].

Firstly an example of the isoluminol system will be given. The mechanism for the chemiluminescence of luminol derivatives is based on the reaction of luminol with hydrogen peroxide and a base catalyst forming an excited solute which emits light. Normally, in CLC the hydrogen peroxide and the catalyst are added post-column to the effluent as two separate solutions, because hydrogen peroxide reacts with the catalyst. However, as mentioned earlier, every additional pumping system will increase the background noise. The inconvenience of handling three flowing solutions (i.e., eluent, hydrogen peroxide and catalyst) can be circumvented by using the electrochemical generation of hydrogen peroxide [34]. This reagent can be generated on-line from the oxygen present in the eluent (Figure 7).

An example of this system is the pre-column derivatisation of the analgesic drug ibuprofen with N-(4-aminobutyl)-N-ethylisoluminol (ABEI). The carboxylic acid function of the analyte is first activated with a carbodiimide, and subsequently reacted with the primary amine function of the isoluminol derivative. The detection limits for ibuprofen in plasma was about 0.15 pmol [34].

The most frequently applied form of chemiluminescence is the peroxyoxalate chemiluminescence, in which an aryl oxalate reacts with hydrogen peroxide forming a reactive dioxetane dione which reacts with a fluorophore resulting in an excited fluorophore which finally emits the light. This means that using this type of chemiluminescence reaction, all fluorophores with an excitation energy of less than 105 kcal/mol can be determined [35].

The combination of chemiluminescence, providing high sensitivity, and immobilised enzyme reactors, providing high specificity, is one of the most powerful reaction-detection systems in CLC. An application is the determination of choline and acetylcholine in serum and urine. The system set-up consists of an eluent pump, a vessel for the eluent, an injection valve, a cation-exchange column for the separation of choline and acetylcholine, and an enzymatic post-column reactor in which acetylcholine esterase and choline oxidase are immobilised on Sepharose (Figure 8). During the enzymatic reaction betaine and hydrogen peroxide are formed in the reactor. The effluent of the reactor is mixed with a make-up flow and pumped through a column containing solid trichlorophenyloxalate and an immobilised fluorophore (aminofluoranthene on glass beads). The make-up flow is needed because the

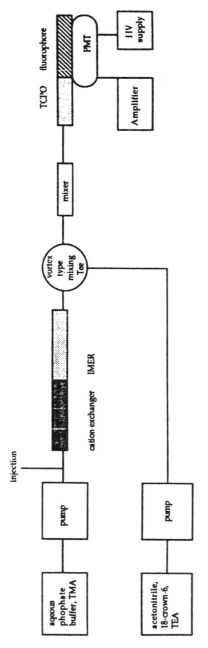

Figure 8.   Schematic of acetylcholine/choline immobilised enzyme reactor system

enzymatic reaction ideally should be performed in pure water, while the chemiluminescence reaction should be performed in an organic solvent. The detection limits were about 500 fmol for choline and acetylcholine with a repeatability of 3.5%. Using this type of immobilised enzyme reactor over 200 samples can be analysed before the reactor needs to be renewed.

CONCLUSIONS

Sample preparation is still the rate-limiting step in many column liquid chromatographic applications. However, these examples of simple chemical manipulation reactions show that if something is known about the physico-chemical properties of the analyte(s), time-consuming and laborious sample clean-up procedures can often be (partly) avoided. If not they can at least be combined with the necessary sensitivity and/or selectivity increasing step. Furthermore, most of these chemical manipulations can be performed on-line and automatedly allowing routine analyses with high accuracy and precision.

From the few illustrated examples of pumpless post-column reaction-detection systems described it will be obvious that, especially in the last decade, these systems have become increasingly popular in biomedical and environmental research areas as a way to improve selectivity and/or sensitivity. Commercial equipment is starting to become more widely available (although less readily than in the area of on-line automated pre-column technology). Additional hardware, such as mixing Tees, phase separators and reactor construction materials, are today of such quality that extra-column band broadening can be kept to a minimum, for both conventional-size and microbore CLC applications.

REFERENCES

1.   R.E. Majors, *LC.GC Int.* 4(2):10-18 (1991).
2.   H. Lingeman, R.D. McDowall, and U.A.Th. Brinkman, *Trends Anal. Chem.* 10:48-59 (1991).
3.   R.D. McDowall, Sample preparation for biomedical analysis. *J. Chromatogr.* 492:3-58 (1989).
4.   J. Chamberlain. Sample preparation by chemical masking. *In:* 'Assay of Drugs and Other Trace Compounds in Biological Fluids', Methodological Surveys in Biochemistry, Vol. 5, E. Reid, ed, Ellis Horwood, Chichester (1976), pp125-131.
5.   M.C. Rouan. Microbore liquid chromatographic determination of cadralazine and cephalexin in plasma with large-volume injection. *J. Chromatogr.* 426: 335-344 (1988).
6.   R.B. Geerdink. Direct determination of metamitron in surface water by large sample volume injection. *J. Chromatogr.* 543: 244-249 (1991).
7.   P.J. Schoenmakers, *J. Chromatogr. Libr.* Vol. 35, Elsevier, Amsterdam (1986), pp20-36.
8.   S. Katz, W.W. Pitt, and G. Jones, *Clin. Chem.* 19:817-819 (1973).
9.   S. Inoue, T. Tokuyama, and K. Takai. Picomole analyses of tryptophan by derivatisation to 9-hydroxymethyl-β-carboline. *Anal. Biochem.* 132:468-480 (1983).
10. I. Jane, and J.F. Taylor. Characterisation and quantitation of morphine in urine using high-pressure liquid chromatography with fluorescence detection. *J. Chromatogr.* 109:37-42 (1975).
11. P.E. Nelson, S.L. Nolan, and K.R. Bedford. High-performance liquid chromatography detection of morphine by fluorescence after post-column derivatisation. *J. Chromatogr.* 234:407-414 (1982).
12. J.P. Langenberg, and U.R. Tjaden. Determination of (endogenous) vitamin K_1 in human plasma by reversed-phase high-performance liquid chromatography using fluorometric detection after post-column electrochemical reduction. *J. Chromatogr.* 305:61-68 (1984).
13. W.F. Bayne, T. East, and D. Dye. High-pressure liquid chromatographic method with postcolumn, in-line hydrolysis and fluorometric detection for indomethacin in biological fluids. *J. Pharm. Sci.* 70:458-459 (1981).
14. M.S. Bernstein, and M.A. Evans. High-performance liquid chromatography-fluorescence analysis for indomethacin and metabolites in biological fluids. *J. Chromatogr.* 229:179-187 (1982).
15. P.J. Twichett, P.L. Williams, and A.C. Moffat. Photochemical detection in high-performance liquid chromatography and its application to cannabinoid analysis. *J. Chromatogr.* 149:683-691 (1978).

16. S.H. Lee, L.R. Field, W.H. Howald, and W.F. Trager. High-performance liquid chromatographic separation and fluorescence detection of warfarin and its metabolites by postcolumn acid/base manipulation. *Anal. Chem.* 53:467-471 (1981).

17. S.P. Assenza, and P.R. Brown. Ultraviolet and fluorescence characterisation of purines and pyrimidines by post-column pH manipulation. *J. Chromatogr.* 289:355-365 (1984).

18. M.S. Gandelman, and J.W. Birks. Liquid chromatographic detection of cardiac glycosides, saccharides and hydrocortisone based on the photoreduction of 2-tert-butylanthraquinone. *Anal. Chim. Acta* 155:159-171 (1983).

19. M.S. Gandelman, and J.W. Birks. Photoreduction-fluorescence detection of aliphatic alcohols, aldehydes, and ethers in liquid chromatography. *Anal. Chem.* 54:2131-2133 (1982).

20. K. Miyazaki, K. Ohtani, K. Sunada, and T. Arita. Determination of ampicillin, amoxicillin, cephalexin, and cephradine in plasma by high-performance liquid chromatography using fluorometric detection. *J. Chromatogr.* 276:478-482 (1983).

21. N. Nimura, K. Ishida, and T. Kinoshita. Novel post-column derivatisation method for the fluorimetric determination of norepinephrine and epinephrine. *J. Chromatogr.* 221:249-255 (1980).

22. G.R. Gotelli, J.H. Wall, P.M. Kabra, and L.J. Marton. Fluorometric liquid-chromatographic determination of serum cortisol. *Clin. Chem.* 27:441-443 (1981).

23. J.C. Gfeller, G. Frey, and R.W. Frei. Post-column derivatisation in high-performance liquid chromatography using the air segmentation principle: application to digitalis glycosides. *J. Chromatogr.* 142:271-281 (1977).

24. U.A.Th. Brinkman, R.W. Frei and H. Lingeman. Post-column reactors for sensitive and selective detection in high-performance liquid chromatography: categorisation and applications. *J. Chromatogr.* 492:251-298 (1989).

25. W.Th. Kok, W.H. Voogt, U.A.Th. Brinkman, and R.W. Frei. On-line electrochemical reagent production for fluorescence detection of phenothiazines in liquid chromatography. *J. Chromatogr.* 354:249-257 (1986).

26. M. Ozcimder, A.J.H. Louter, H. Lingeman, W.H. Voogt, R.W. Frei, and M. Bloemendaal. Determination of oxidized, reduced and protein-bound glutathione in eye lenses by high-performance liquid chromatography and electrochemical detection. *J. Chromatogr.* 570:19-33 (1991).

27. H. Jansen, C.J.M. Vermunt, U.A.Th. Brinkman, and R.W. Frei. Parallel column ion exchange for post-separation pH modification in liquid chromatography. *J. Chromatogr.* 366:135-144 (1986).

28. H. Jansen, R. Jansen, U.A.Th. Brinkman, and R.W. Frei. Fluorescence enhancement for aflatoxins in HPLC by post-column split-flow iodine addition from a solid-phase reservoir. *Chromatographia* 24:555-559 (1987).

29. J.W. Birks, and R.W. Frei. Photo-chemical reaction detection in HPLC. *Trends Anal. Chem.* 1:361-367 (1982).

30. C. de Ruiter, J.F. Bohle, G.J. de Jong, U.A.Th. Brinkman and R.W. Frei. Enhanced fluorescence detection of dansyl derivatives of phenolic compounds using a postcolumn photochemical reactor and application to chlorophenols in river water. *Anal. Chem.* 60:666-670 (1988).

31. C.E. Werkhoven-Goewie, W.M.A. Niessen, U.A.Th. Brinkman, and R.W. Frei. Liquid chromatographic detector for organosulphur compounds based on a ligand-exchange reaction. *J. Chromatogr.* 203:165-172 (1981).

32. H. Irth, G.J. de Jong, U.A.Th. Brinkman, and R.W. Frei. Metallic copper-containing post-column reactor for the detection of thiram and disulfiram in liquid chromatography. *J. Chromatogr.* 370:439-447 (1986).

33. K. Imai, *in*: 'Detection-Oriented Derivatization Techniques in Liquid Chromatography', chromatographic Science Series, Vol. 48, H. Lingeman and W.J.M. Underberg, eds., Dekker, New York (1990).

34. O.M. Steijger, G.J. de Jong, J.J.M. Holthuis and U.A.Th. Brinkman. On-line electrochemical reagent generation for liquid chromatography with luminol-based chemiluminenscence. *J. Chromatogr.* 557:13-21 (1991).

35. P. van Zoonen, C. Gooijer, N.H. Velthorst, R.W. Frei, J.H. Wolf, J. Gerrits and F. Flentge. HPLC detection of choline and acetylcholine in serum and urine by an immobilised enzyme reactor followed by chemiluminescence detection. *J. Pharm. Biomed. Anal.* 5:485-492 (1987).

PROTON NUCLEAR MAGNETIC SPECTROSCOPY: A NOVEL METHOD FOR THE STUDY OF SOLID PHASE EXTRACTION

I.D. Wilson[1] and J.K. Nicholson[2]

[1] Department of Safety of Medicines
ICI Pharmaceuticals
Mereside
Alderley Park
Macclesfield
Cheshire, SK10 4TG, UK

[2] Department of Chemistry
Birkbeck College
Gordon House
29 Gordon Square
London,
WC1E 0PP, UK

SUMMARY

The use of proton (^{1}H) NMR as a method for studying the solid phase extraction (SPE) of biofluids is described with particular reference to studies undertaken on human urine. The extraction of both control urine and samples containing ibuprofen or paracetamol metabolites was used to investigate the extraction properties of a variety of phases for both endogenous compounds and drug metabolites. The phases studied included C_{18}, β-cyclodextrin, aminopropyl, strong anion and strong cation exchange materials (SAX and SCX) and a multimodal phase exhibiting reversed-phase, SAX and SCX properties. In addition to enabling the study of the extraction properties of the sample components NMR also revealed the presence of phase-related material in the eluates from some SPE phrases.

The advantages of ^{1}H NMR as a multiparametric "universal" detector for this type of investigation are contrasted with other general means of detection such as UV spectroscopy.

INTRODUCTION

There are probably as many approaches to the development of solid phase extraction (SPE) methods as there are laboratories using this methodology. Clearly however, at least four variables must be considered when devising SPE methods. These are the nature of the analyte; the type of matrix; the SPE phase to be used; and the method used to analyse the prepared sample (the analytical "end step"). In general, sample preparation can be considered a combination of "sample clean up" (i.e. the selective removal of interferences) and concentration of the analyte in order to provide sufficient of the compound of interest to ensure reliable detection and quantification. Obviously the balance between the clean up and concentration required will be greatly affected by the analytical end step employed. Thus,

Sample Preparation for Biomedical and Environmental Analysis,
Edited by D. Stevenson and I.D. Wilson, Plenum Press, New York, 1994

37

for methods where the analytical end step is both sensitive and specific (i.e. chromatography linked to mass spectrometry) minimal sample preparation may be required. However, significantly more may be required, of both clean up and preconcentration steps, for the assay of the same analyte by a less sensitive and specific method (e.g. high performance liquid chromatography (HPLC) with UV detection at 254 nm). In the case of a non-specific assay the selective removal of interferences represents a key step, and it is obvious that a knowledge of the identity of any co-extracting compounds would be of inestimable value. Such information would enable rational methods for the removal of contaminants during sample preparation to be devised, rather than the analyst having to rely on more pragmatic approaches (i.e. "trial and error"). It is the case that the chemical composition of most common biofluids, including plasma, urine bile, milk, etc, is generally well characterised. However, whilst the composition of these biofluids may be known the extraction properties of many of their constituents has not been determined. In order to investigate the nature of co-extracted interferences the analytical requirement is for a general method for detecting all of the components in the sample and in the subsequent extract. However, the method must also enable the identity of the various components to be determined. Proton (^{1}H) NMR spectroscopy has many of the desirable attributes of such a detector, and indeed we have used it as such in the study of toxicology and xenobiotic metabolism (1-6). Here we describe the extraction of endogenous compounds together with a number of model drugs and their metabolites from urine, onto a variety of SPE cartridges, monitored by ^{1}H NMR. The use of NMR spectroscopy in this way not only allowed the extraction of the compounds of interest to be followed, but also enabled bulk co-extracted endogenous components, together with material leaching from the SPE phase itself, to be detected and identified.

MATERIALS AND METHODS

Samples

Urine was obtained from a male volunteer before dosing, and following the administration of a single therapeutic dose of either paracetamol (acetaminophen) (500 mg) or ibuprofen (400 mg) for the periods 0-2 hr post dose. Samples were stored frozen at -20°C until analysed as described below.

UV Spectroscopy

UV spectra were obtained on a Phillips PU 8720 UV/Visible spectrophotometer between 230 and 450 nm. UV Spectra of samples and appropriate extracts were obtained in aqueous (water or D_2O) solution in 1 cm quartz cuvettes.

NMR Spectroscopy

^{1}H NMR spectroscopy was performed on freeze dried samples reconstituted in D_2O using a field strength of 250 MHz on a Bruker 250 AM NMR spectrometer. Spectra were the result of 64 free induction decays (Fids) which were collected into 32 x 1024 data points. A gated secondary irradiation was applied at the water resonance frequency in order to suppress the signal from residual water protons.

Solid Phase Extraction

Solid phase extraction was performed using C_{18}, SAX, SCX, aminopropyl and mixed mode Bond Elut (JM) cartridges (Varian Associates, purchased from Jones Chromatography, Hengoed, UK), and on Spe-ed (JM) Cyclobond 1 SPE cartridges (Applied Separations, PA 18014, USA).

For extraction onto the 1 ml C_{18} bonded silica gel cartridges 2 ml of sample was taken and acidified to pH 2 with 0.1 M HCl. The cartridges were activated prior to sample application by washing with methanol (1 ml) and acidified water (pH 2 with HCl, 1 ml). After sample application the cartridges were washed with acidified water (0.3 ml) prior to elution with methanol (2 ml). In the case of the 6 ml cartridges a water wash of 1 ml and elution with 4 ml of methanol was adopted. For some experiments stepwise gradient elution with 20, 40, 60, 80 and 100% acidified methanol-water mixtures was employed (see text). Essentially the same protocol was used for the Cyclobond 1 cartridges as described for the C_{18} bonded materials above.

In the case of the ion-exchange materials (aminopropyl, SAX, SCX, mixed mode) the cartridges were washed with methanol (2 ml) and water (pH 7, 2 ml) prior to the application of the sample (2 ml, pH 7). Following sample application the cartridges were washed with water (1 ml, pH 7). Aminopropyl and SAX cartridges were then eluted with methanolic-1M HCl (3 ml 95:5 V/V). The SCX cartridges were eluted with methanol-aqueous ammonia (3 ml sg. 0.88, 95:5 V/V). Elution from the mixed mode cartridges was accomplished by elution first with methanolic-HCl and then with methanol-aqueous ammonia as described for the SAX and SCX cartridges above.

The various eluates were prepared for spectroscopy by removing as much methanol as possible using a stream of nitrogen followed by freeze drying. The residues were then taken up in 1 ml of D_2O (99.9 atom% D, Aldrich Chemical Co Ltd, Gillingham, UK) and transferred to NMR tubes for spectroscopy.

RESULTS AND DISCUSSION

The value of NMR spectroscopy in the context of the study of SPE (or, for that matter, any type of sample preparation) lies in the ability of the technique to provide a general means of detection whilst at the same time providing a wealth of structural information to enable identification. This is perhaps best demonstrated by comparison with UV detection which also provides a fairly general means of detection but little structural information. Thus, in Figure 1a-c the UV and ^{1}H NMR spectra of three urine samples are shown. These show a control human urine sample (1a), a urine obtained from a human volunteer for the period 0-2 hr after a 400 mg/ oral dose of ibuprofen (1b) and a similar urine sample obtained for the period 0-2 hr after oral administration of 500 mg of paracetamol (1c). Whilst the UV spectra for all three samples do show small differences it is clear that the information content is low and certainly would not allow for the unambiguous identification of any of the constituents. In contrast, the three NMR spectra display a wealth of detail concerning both endogenous compounds and the excreted drugs and their metabolites. Indeed major differences are apparent between the control urine and the samples obtained following dosing with the two drugs. Certainly all three spectra have quite distinct "fingerprints" and those containing drug metabolites can not only be distinguished from the control but also from each other. In these samples it is possible to identify resonances from endogenous components such as hippuric acid, creatinine, citrate, dimethylamine, succinate and α-ketoglutarate (Figure 1a-c). In the case of the ibuprofen metabolite-containing urine a broad envelope of signals can also be seen in the aromatic region of the spectrum, as well as prominent signals in the aliphatic portion of the spectrum corresponding to the various methyl groups present on the analytes (see reference 3 for signal assignments). Similarly with the paracetamol metabolite-containing urine prominent signals are present for the sulphate and glucuronide metabolites, both for the aromatic protons and the N-acetyl resonances (see reference 3 for signal assignments).

Having characterised the samples using ^{1}H NMR these urines were then used to probe the various SPE phases in order to determine their extraction properties.

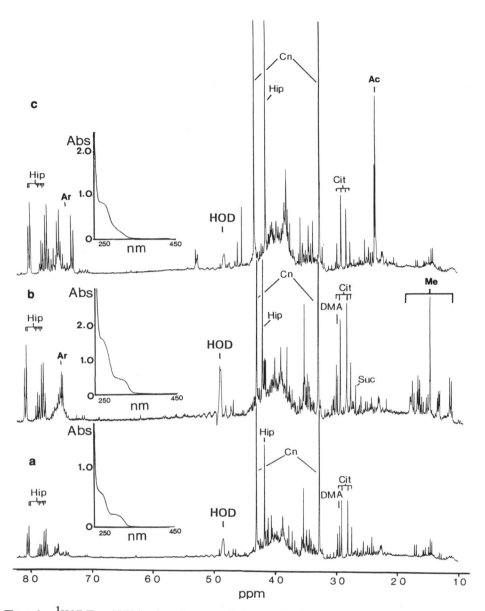

Figure 1. ^{1}H NMR and UV (see insets) spectra of urine samples, freeze dried and reconstituted in D$_2$0 for analysis, (a) control human urine; (b) human urine following the administration of 400 mg of ibuprofen to a normal male volunteer (0-2 hr sample); (c) human urine following the administration of 500 mg of paracetamol to a normal male volunteer (0-2 hr sample).

KEY: Cit, citrate; Cn, creatinine; DMA, dimethylamine; Hip, hippurate; HOD, residual water; succ, succinate; Ar, aromatic protons; Ac, acetyl protons; Me, methyl protons.

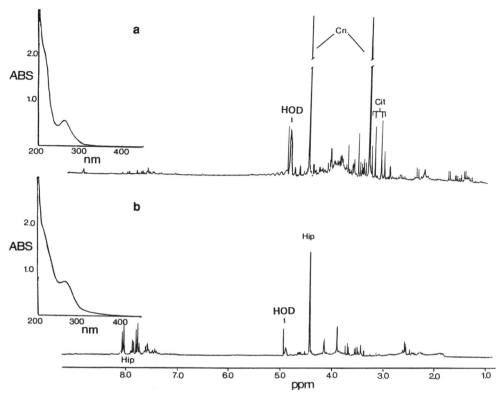

Figure 2. ^{1}H NMR and UV (see insets) spectra for control human urine (see Figure 1) following solid phase extraction on a 3 ml C$_{18}$ Bond Elut cartridge as described in the text. (a) non-retained material; (b) methanolic eluate.
KEY: as for Figure 1.

Extraction Of Urine Using A C-18 Bonded SPE Phase

Initial experiments on the SPE of urine involved the use of C$_{18}$ bonded silica gel. Acidified control urine (pH 2, 2 ml) was passed through a cartridge (100 mg) previously conditioned with acidified methanol and water. The non-retained portion and the material eluting with a methanol wash were then taken, blown to dryness and/or freeze dried and redissolved in D$_2$O for analysis by UV and NMR spectroscopy. The results of this experiment are illustrated in figure 2 a and b. As can be seen the UV spectra for both fractions are very similar and relatively uninformative. However, the NMR spectra are quite revealing and show, for example, that creatinine and citrate are completely unretained under the conditions employed here. In contrast hippurate, a major endogenous urinary component, is well retained and elutes with the methanol wash. In Figure 3a the spectrum obtained for the methanolic eluate for a similar experiment performed with 2 ml of acidified urine containing ibuprofen metabolites, which employed 500 mg of sorbent, is shown. Comparison of this spectrum with that of the original sample (Figure 1b) shows that essentially all of the drug-related material has been extracted, together with hippurate, and perhaps more surprisingly citrate and some creatinine (neither of which were retained from the control urine using 100 mg of sorbent). In fact this spectrum graphically demonstrates the need for careful optimisation of the amount of sorbent used for extraction. In this case the 500 mg of adsorbent used

41

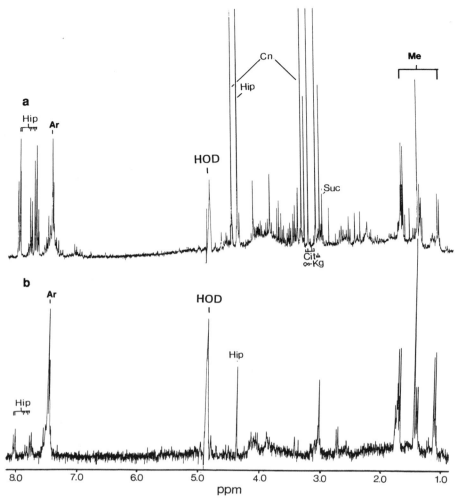

Figure 3. ^{1}H NMR spectra of extracts of ibuprofen metabolite-containing urine demonstrating the effect of changing the ratio of sorbent to sample on the composition of the final methanolic eluate from C_{18} Bond Elut cartridges. (a) methanolic eluate from a 500 mg cartridge; (b) methanolic eluate from a 25 mg cartridge.
KEY: as for Figure 1.

represents a huge excess of material, and large quantities of endogenous compounds have been coextracted with the ibuprofen metabolites. Investigation of the optimum amount of C_{18} bonded sorbent required for complete extraction of only the ibuprofen-related components revealed that as little 25 mg of sorbent was sufficient. The resulting NMR spectrum is illustrated in Figure 3b where it can be seen, in marked contrast to the result obtained for 500 mg of sorbent, that apart from the ibuprofen-related material the only other signals result from the presence of a small amount of hippurate. Clearly the temptation to use an excess of sorbent to ensure high extraction efficiencies thus carries with it the attendant risk of a lower degree of sample clean up. This will be especially the case when well retained analytes, such as ibuprofen and its metabolites, are being extracted. Such a result may also explain why less retentive phases (e.g. C-2 or C-8) are reported to give "cleaner" extracts compared to C-

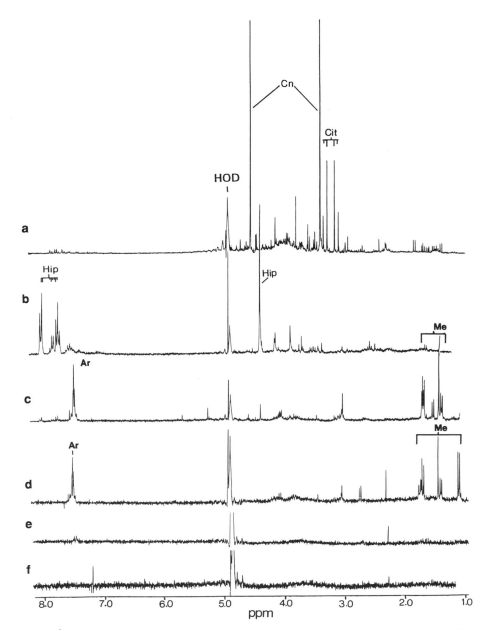

Figure 4. ^{1}H NMR spectra of the eluates from a C_{18} Bond Elut cartridge following the application of 2 ml of ibuprofen metabolite-containing urine and stepwise gradient elution with methanol-acidified water (pH 2).   (a) non-retained material; (b) methanol-water 20:80; (c) methanol-water 40:60; (d) methanol-water 60:40; (e) methanol-water 80:20; (f) methanol.
KEY: as for Figure 1.

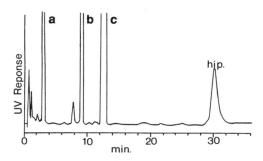

Figure 5. Supercritical fluid chromatogram of the methanolic eluate obtained from a C_{18} SPE cartridge following the application of an acidified urine sample that had been spiked with 3 acidic drugs (peaks a,b and c). The late running peak was identified as hippuric acid (hip) based on a consideration of the likely endogenous contaminants resulting from such an extraction scheme and subsequent co-chromatography with an authentic standard.

18 bonded silica when compared on a mg for mg of adsorbent basis. The use of 15 and 20 mg of sorbent was also investigated to see if the small amount of residual hippurate still contaminating the extract could be eliminated by the use of even smaller amounts of C-18 bonded silica. However, we experienced severe problems in packing such small amounts of material into the cartridges. The observed breakthrough of some of the ibuprofen metabolites with the 15 and 20 mg cartridges may therefore have been due more to uneven packing rather than a lack of capacity.

The alternative strategy, and the one most widely adopted by analysts, is of course to use an intermediate wash in order to eliminate more polar, less well retained, contaminants. The result of washing the cartridges with a 20:80 mixture of methanol-water (pH 2) was the complete elution of any remaining hippuric acid, without loss of the ibuprofen metabolites. The ibuprofen related material was recovered in a 60:40 methanol-water (pH 2) wash. Alternatively, a stepwise gradient of increasing eluotropic strength may be employed in order to partially fractionate the metabolites as illustrated in figure 4 a-f. As can be seen from this figure essentially all of the endogenous materials were either not retained or eluted in the first 20% methanolic wash step. The ibuprofen related material was eluted in the 40:60 and 60:40 methanol-water (pH 2) washes. The 40:60 fraction contained mainly the 2-(4-(2-hydroxy-2-methylpropyl)phenyl)propionic acid and 2-(4-(2-carboxy propyl)phenyl) propionic acid metabolites (signal assignments in reference 3) whilst the 60:40 wash was enriched in ibuprofen. Clearly the subfractionation of ibuprofen metabolites illustrated here is by no means complete. However, as we have described in detail elsewhere, this methodology can be used to obtain milligram quantities of essentially pure metabolties providing that the metabolite profile is not too complex (3-6). This is essentially a simple, low resolution, form of chromatography. We have termed this procedure Solid Phase Extraction/Chromatography (SPEC)-NMR.

One instance where a knowledge of the extraction/elution profile of the endogenous urinary components proved to be of practical benefit is illustrated in figure 5. This shows the result of the supercritical fluid chromatography (SFC) of the methanolic eluate obtained for a urine sample, spiked with three acidic drugs, following extraction at pH 2 onto a C_{18} bonded SPE cartridge. Clearly, in addition to the three compounds of interest there is also a late running endogenous contaminant. Based on the results illustrated in figures 2 to 4 we reasoned that this contaminating material was most likely to be hippuric acid. Hippuric acid did indeed have the same chromatographic retention time as the unknown peak. Based on our knowledge of the extraction/elution profile of hippuric acid we therefore inserted an

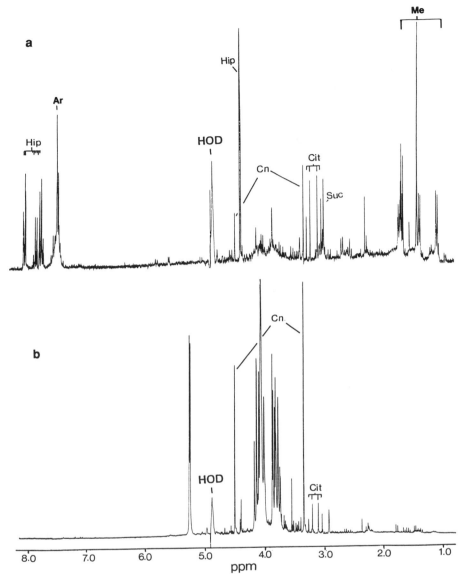

Figure 6. ^{1}H NMR spectra of eluates from a Cyclobond 1 SPE cartridge following the application of 2 ml of an ibuprofen metabolite-containing urine (pH 2). (a) non-retained material together with phase-related contaminant and (b) methanolic eluate.
KEY: as for Figure 1.

intermediate methanol-water (pH 2) 20:80 wash step prior to the elution of the analytes with 100% methanol. This eliminated the late running peak and allowed us to reduce the analysis time by ca. 50% (7). No doubt we could have arrived at the same solution by trial and error but somewhat less efficiently.

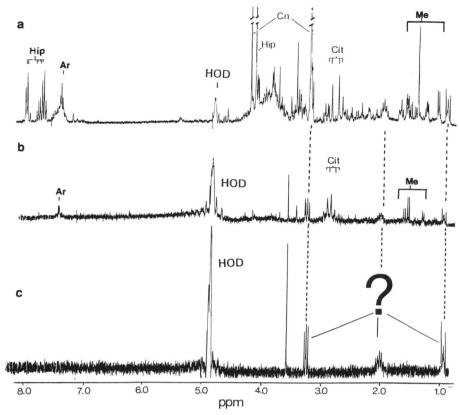

Figure 7. ^{1}H NMR spectra of eluates from a 100 mg aminopropyl SPE cartridge following the application of 2 ml of an ibuprofen metabolite-containing urine (pH 7). (a) non-retained material (b) methanolic eluate and (c) combined methanol-water conditioning washes showing the presence of phase-relate ted material. This material can also be seen as a contaminant of the eluates shown in spectra a and b.   KEY: as for Figure 1.

Extraction Of Urine Using A β-cyclodextrin Bonded SPE Phase

The C_{18} bonded phase is a typical representative of the "reversed-phase" SPE mate-rials available, and C-2 and C-8 materials in general give similar results, albeit with less eluotropic solvent mixtures for the same elution profiles. The cyclodextrin bonded phase in contrast is based on the extraction of the analyte into the cavity of the cyclic oligosaccahride β-cyclodextrin bonded to a silica gel substrate. The results obtained for the extraction of 2 ml of ibuprofen metabolite-containing urine on a 500 mg Cyclobond 1 cartridge are shown in Figure 6. It is immediately apparent that the non-retained fraction (Figure 6a) contained, in addition to creatinine and citrate, considerable amounts of material not present in the original sample. This material was clearly derived from the β-cyclodextrin used to prepare the phase. However, extensive washing of the cartridge did remove this phase-related mate-rial and when the cartridge was eluted with methanol the resulting extract was free of con-tamination. The material in the methanolic eluate included hippurate, the ibuprofen-related material, some citrate and a small amount of creatinine (Figure 6b). These results were therefore similar to those obtained for the 500 mg C-18 phase described above.

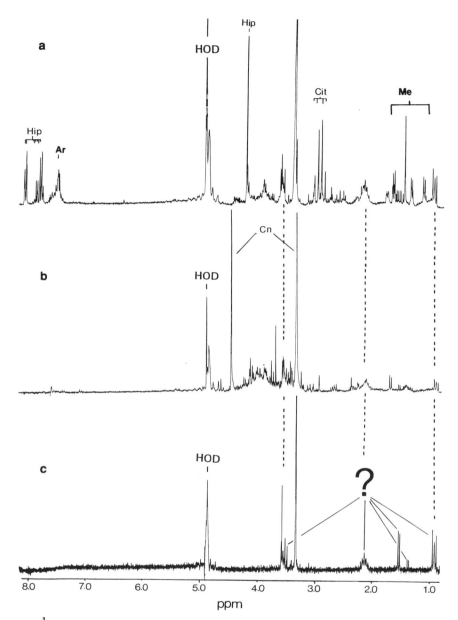

Figure 8. ^{1}H NMR spectra of eluates from a 500 mg SCX cartridge following application of 2 ml of an ibu profen metabolite-containing urine (pH 7). (a) acidified methanolic eluate and (b) non-retained material. Spectrum c shows the phase-related material obtained by washing the column with methanol. This material can also be seen as a contaminant of the eluates shown in spectra a and b. KEY: as for Figure 1.

Extraction Of Urine Using An Aminopropyl Bonded SPE Phase

Whilst the hydrophobic interaction mechanisms used for extraction onto the C-18 and cyclodextrin bonded phase, combined with ion-suppression in the case of ionisable analytes, represent one means of sample preparation the use of ion exchange phases is another option. The aminopropyl bonded phase can function as a weak ion-exchange material suitable for the extraction of acids. The extraction of the acidic ibuprofen metabolites onto the amino-propyl bonded silica gel was investigated by applying 2 ml of the test urine (pH 7) onto a 100 mg cartridge. As can be seen from Figure 7b the bulk of the components of the sample were not retained on the phase under the conditions used. This should perhaps be contrasted with the excellent extraction achieved with a similar quantity of the C-18 bonded material. The selectivity of the phase was however, somewhat different to the C-18 material in that some citrate was extracted together with a small amount of ibuprofen (Figure 7a). This is interesting given that the polarity of citrate was such that it was completely unretained on the C-18 bonded phase under conditions that resulted in the complete retention of all the ibuprofen related material. Another interesting feature of these spectra is the presence of additional resonances in some of them, which were not present in the original urine NMR spectrum. As observed with the cyclodextrin-bonded phase it would appear that material derived from the SPE phase was eluted from the column together with the analytes and matrix components. This was confirmed by collecting the methanol and water washes used for activating the column for subsequent analysis by NMR. Thus the spectrum shown in Figure 7c clearly show the resonances for the contaminant detected in Figure 7a and b.

Extraction Of Urine Using A Strong Anion Exchange SPE Phase

As a result of our experiments using the aminopropyl phase we subsequently examined the use of SPE cartridges containing a strong anion exchanger (SAX). Using cartridges containing 100 mg of material we observed considerable breakthrough albeit, less than that seen with the amino bonded phase. Elution with methanolic HCl resulted in the recovery of quantities of hippurate and ibuprofen metabolites and we therefore repeated the extraction using cartridges containing more of the SAX phase. The use of cartridges containing 500 mg of material enabled the extraction of essentially all of ibuprofen related material together with all of the acidic or amphoteric endogenous metabolites, including citrate (Figure 8a). The extracted material was recovered in a single step using methanolic HCl. As would be expected basic compounds such as creatinine and trimethylamine N-oxide were not retained by the SAX phase (Figure 8b). As noted with the aminopropyl bonded phase we also observed new resonances in the spectra resulting from the presence of phase related-material. Interestingly this phase related material was not removed from the SPE cartridge by washing with water but was eluted with methanol (Figure 8c). We have not observed this phase-related material with all batches of cartridges. It did however, seem to be more apparent with older batches of cartridges, suggesting either a change in the manufacturing process or some slow decomposition due to the "ageing" of the phase.

Extraction Of Urine Using A Strong Cation Exchange SPE Phase

Having examined the use of aminopropyl and SAX phases for the extraction of urine containing ibuprofen metabolites we then investigated the use of the strong cation exchange material on the same samples. The results for the extraction of 2 ml of ibuprofen metabolite-containing urine (pH 2) onto a 500 mg SCX cartridge are illustrated in Figure 9. Essentially all of the acidic ibuprofen metabolites and endogenous compounds passed through the cartridge unretained (Figure 9a). The retained material was eluted with ammoniacal methanol,

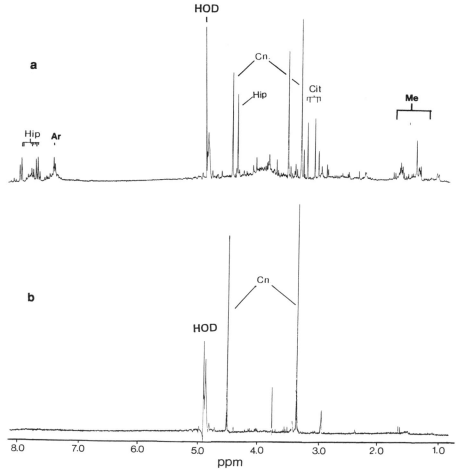

Figure 9. ^{1}H NMR spectra of the eluates from a 100 mg SCX cartridge following application of 2 ml of ibu profen metabolite-containing urine (pH 7). (a) non-retained material and (b) methanol-ammonia eluates.
KEY: as for Figure 1.

and consisted mainly of creatinine together with small amounts of betaine and dimethyla-mine (Figure 9b). The use of cartridges containing 100 mg of the SCX phase gave similar results but with some breakthrough of the creatinine.

Once again NMR spectra of extracts from some batches of cartridges revealed the presence of resonances that were not present in the original sample. An example of this is shown in figure 10 where the methanolic eluate from a 500 mg SCX cartridge is shown. The characteristic signals for a para-disubstituted benzene ring are clearly visible. The SCX phase is based on a long chain benzene sulphonic acid bonded to silica gel. The inevitable conclusion therefore must be that, as seen with the cyclodextrin, amino and SCX phases described above, that these signals are due to leaching of the ion exchange functionality from the phase following hydrolysis of the bond linking it to the silica substrate. As seen with the SAX phase washing the cartridge with water did not elute any of this phase-related material.

49

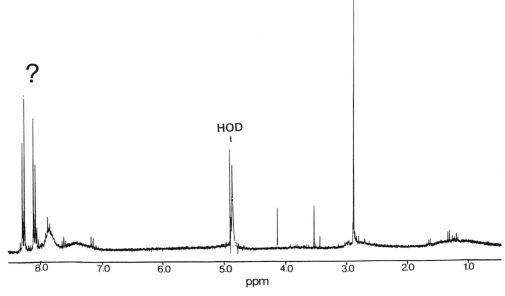

Figure 10. ^{1}H NMR spectrum of the methanol wash of a 500 mg SAX cartridge showing the phase-related material eluted under these conditions.

However, washing with methanol resulted in the "recovery" of significant quantities of this material (Figure 10). As with the SAX phase this material was not seen in the methanolic eluates from all batches of cartridges.

Extraction Of Urine Using A Multimodal SPE Phase

As illustrated in the examples provided above, by selection of the appropriate SPE phase the extraction of any of the components of the ibuprofen-metabolite containing urine is possible. However, no phase appears to be capable of providing a means for extracting all the sample components under a single set of conditions. This is hardly unexpected given the mixture of acidic, basic and amphoteric compounds, covering a range of polarities, present in the sample. This would require more than one extraction mechanism to be operating simultaneously (i.e. anion and cation exchange) in order to obtain extraction. However, there are circumstances, particularly in drug metabolism studies, where the initial requirement is for the extraction and concentration of metabolites of unknown structure, and therefore unknown extraction properties. A multimodal phase, capable of extracting acidic, basic and neutral compounds would be of considerable value in providing this initial extract. Recently a number of phases have become available which combine reversed-phase and either anion or cation exchange properties (e.g. the Certify range from Varian Assoc.). We have recently investigated the properties of an experimental phase (Varian Assoc.) which combines anion, cation and reversed-phase characteristics. To evaluate this phase we mixed equal volumes of urine containing both paracetamol and ibuprofen metabolites to prepare a sample containing a mixture of endogenous compounds together with a range of drug related material (including both phase 1 and phase 2 metabolites).

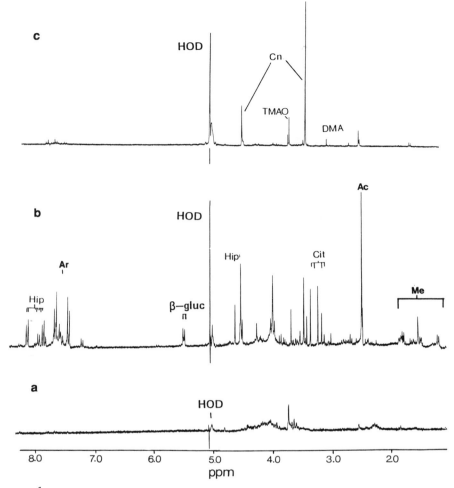

Figure 11. ^{1}H NMR spectra of eluates from a 1500 mg "multimodal" SPE cartridge following application of 2 ml an ibuprofen/paracetamol metabolite-containing urine (pH 7). (a) non-retained material; (b) methanol-HCl and (c) methanol-ammonia eluates.
KEY: as for Figure 1.

The results of the extraction, at neutral pH, of this sample are shown in Figure 11a-c. As can be seen the NMR spectrum of the non-retained material is essentially featureless (Figure 11a) showing the complete extraction of both endogenous compounds and drug metabolites. Elution with methanolic HCl resulted in the spectrum shown in Figure 11b. This spectrum contains resonances for hippuric acid, citrate, etc, together with the ibuprofen and paracetamol-related material. Subsequent elution with ammoniacal methanol resulted in the recovery of creatinine and other basic compounds (Figure 11c). Thus despite providing a general and non-specific means of extraction this type of SPE column still enabled the selective elution of compounds by class. We have used this material to advantage as a means of extracting an "unknown" endogenous substance that appeared in large quantities in the urine of rats dosed with paracetamol for extended periods. The initial extraction and concentration of this material using this multimodal phase was possible without having to make any assumptions about its structure or extraction properties. Following its recovery from the column NMR and mass spectrometry identified the compound as 5-oxoproline (MacFarlane *et al*, in preparation).

CONCLUSIONS

As we have shown ^{1}H NMR provides a rapid, simple and convenient method for the multicomponent assay of both biofluids and the eluates from SPE columns. A clear lesson from these studies with respect to the C-18 bonded phase is that often much less material is required to give good extraction of well retained material, like the ibuprofen metabolites, than might be thought, and that the use of an "excess" of the phase is counterproductive as it merely results in the extraction of additional unwanted endogenous contaminants. Another useful benefit of the use of NMR to analyse these extracts was the ability of the technique to detect contamination of the extract by the SPE phase used for sample preparation. Whilst such contamination may be of little importance for most applications there is clearly the possibility that material leaking from ion-exchange materials could, over a period of time, modify the behaviour of a chromatographic separation due to the gradual build up of material in the HPLC column (there is much anecdotal evidence that the use of ion-pair reagents results in changes in the subsequent behaviour of RP-HPLC columns). The NMR spectra also served to highlight differences in the selectivity of extraction between the properties of the various phases for the endogenous components present in urine. This has enabled us to devise rapid, rational, methods for the purification and identification of xenobiotic metabolites (4-6). We are now extending these studies to include other important biofluids such as bile, milk and plasma.

REFERENCES

1.   J.K. Nicholson and I.D. Wilson. High resolution proton magnetic resonance spectroscopy of biological fluids. *Prog. NMR Spec.*, 21, 449-501 (1989).
2.   I.D. Wilson, J.K. Nicholson, F.Y.K. Ghauri and C.A. Blackledge. Use of high-field nuclear magnetic resonance spectroscopy for the analysis of biological fluids. *Anal. Proc.*, 28, 217-223 (1991).
3.   I.D. Wilson and J.K. Nicholson. Solid phase extraction chromatography and nuclear magnetic resonance spectrometry for the identification and isolation of drug metabolites. *Anal. Chem.*, 59, 2830-2832 (1987).
4.   I.D. Wilson and I.M. Ismail. A rapid method for the isolation and identification of drug metabolites from human urine using solid phase extraction and proton NMR spectroscopy. *J. Pharm. Biomed. Anal.*, 4, 663-665 (1986).
5.   I.D. Wilson. Solid phase extraction chromatography and NMR spectroscopy (SPEC-NMR) for the rapid identification of drug metabolites in urine. *J. Pharm. Biomed. Anal.*, 6, 151-165 (1989).
6.   K.E. Wade, I.D. Wilson, J.A. Troke and J.K. Nicholson. 19F and 1H magnetic resonance strategies for metabolic studies on fluorinated xenobiotics: Application to flurbiprofen (2-(2-fluoro-4-biphenyl)propionic acid). *J. Pharm. Biomed. Anal.*, 8, 401-410 (1990).
7.   D.W. Roberts and I.D. Wilson. Bioanalytical supercritical fluid chromatography. In: E. Reid and I.D. Wilson (eds) "Analysis for drugs and metabolites", Royal Society of Chemistry, p.257-264 (1990).

THE EFFECT OF CARTRIDGE CONDITIONING IN THE REVERSED-PHASE

EXTRACTION OF BASIC DRUGS

B. Law

ICI Pharmaceuticals
Mereside
Alderley Park
Macclesfield, SK10 4TG. UK

SUMMARY

The blocking of residual silanols on reversed-phase extraction cartridges by preconditioning the cartridges with various cations prior to the application of the basic drugs propranolol and atenolol has been studied. The ability of cations to eliminate secondary interactions through blocking of the silanols was related to their displacement strength as used in cation-exchange chromatography. To enable the cartridges to behave in a true reversed-phase manner and thereby achieve quantitative elution of these basic solutes with methanol or aqueous methanol, high concentrations of inorganic cations, e.g. Na^+, K^+ equivalent to 1ml of a 1M solution were required to condition a 100mg cartridge.

INTRODUCTION

Solid-phase extraction procedures have become well established in the field of drug analysis particularly for basic compounds [e.g. 1-3]. The involvement of secondary interactions on nominally reversed-phase extraction cartridges with this type of compound is now well accepted although probably less well understood. To obtain elution from these "mixed mode" cartridges it is necessary to use some form of competing cation or extremes of pH in the final eluent to overcome the secondary electrostatic interactions [4]. It is these secondary interactions that give reversed-phase cartridges their unique character. Although this mixed mode of extraction has been exploited successfully by a number of workers for the analysis of basic drugs [4-6], it can be a complicating factor in the development of analytical procedures. It was considered of interest therefore to examine the effect of conditioning the cartridge with various cations prior to sample application with a view to eliminating the secondary electrostatic interactions and so producing a simpler system based purely on reversed-phase interactions.

Sample Preparation for Biomedical and Environmental Analysis,
Edited by D. Stevenson and I.D. Wilson, Plenum Press, New York, 1994

53

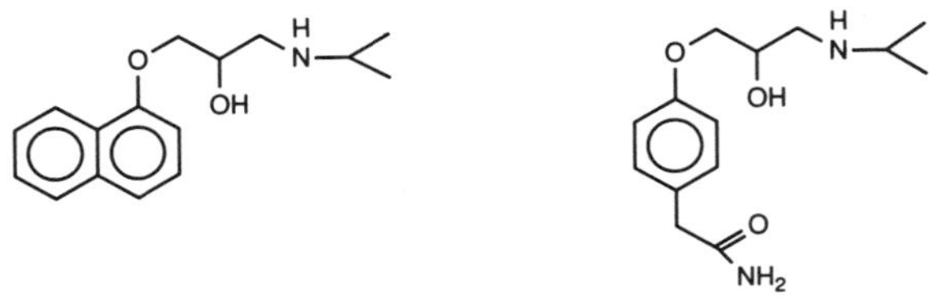

Propranolol   Atenolol

Figure 1.   Structures of propranolol and atenolol.

EXPERIMENTAL

Materials and Equipment

Bond-Elut cartridges, 100mg size (C2 and C18) and a Vac-Elut manifold were obtained from Jones Chromatography, Hengoed, UK. Methanol was HPLC grade from Fisons, Loughbrough, UK, tetrabutylammonium dihydrogen phosphate (97%) was from Aldrich, Gillingham, UK, all other chemicals were Analar grade from E Merck, Liverpool, UK.

The solutes studied, propranolol and atenolol (Figure 1) obtained from the Radio-chemistry Laboratory at ICI Pharmaceuticals, were radiolabelled with ^{14}C at a specific activity of 13.9 and 17.6μCi/mg respectively, and 98% radiochemical purity. Scintillation counting was carried out using a Canberra Packard 1900A TRICARB counter with inbuilt quench correction. The scintillation fluid was Ready Protein Plus from Beckman.

Method

Compounds were applied to the cartridges in water at a concentration of 0.9 to 2.0μg/ml, equivalent to a radioactive concentration of approximately 50×10^3 disintegrations per minute (dpm)/ml. All cartridges were conditioned with methanol (1ml), followed by a salt solution (1ml) in either water, methanol/water (1:1) or methanol and then water (1ml) before use. Acetate salts of the following cations: H^+, Li^+, NH_4^+, Na^+, K^+, Pb^{2+}, TEA^+ (triethylamine) were used for conditionings with the addition of TBA^+ (tetrabutylammonium) which was in the dihydrogen phosphate form. The vacuum to each cartridge was individually controlled with a small stopcock to prevent the cartridge drying prior to sample application. A vacuum pressure of 34kPa (equivalent to a flow rate of approximately 10ml/min for methanol) was used throughout.

A typical experiment involved conditioning the cartridge (with or without a salt solution), application of drug solution (1ml) then sequential elution with a range of elution solvents (1ml), consisting of water and methanol, alone and as mixtures. The application volume along with the elution solvents were collected into polypropylene vials, scintillation fluid added, and the radioactivity eluted at the various washing stages determined. The data was then processed by plotting the cumulative percentage of compound eluted against the stage in the elution process. All experiments were carried out in duplicate and results presented are mean data.

RESULTS AND DISCUSSION

When reversed-phase cartridges are conditioned in the standard way with methanol and water they rarely behave in a truly reversed-phase manner with respect to basic drugs.

Table 1. Recovery of propranolol and atenolol from reversed
phase cartridges following washing with increasing
concentrations of methanol in water.

	C2		C18	
	Atenolol	Propranolol	Atenolol	Propranolol
Cumulative Recovery (%) Over 6 washes	0.27	2.21	6.31	2.38

For example, neither propranolol nor atenolol can be eluted from C2 or C18 cartridges when washed with methanol or 50% aqueous methanol alone (Table 1). This unexpected retention is due to a secondary electrostatic interaction between the basic amine and residual ionised silanols on the surface of the silica. By treating the cartridge with salt solutions prior to sample application it was postulated that the electrostatic interaction could be reduced or eliminated so allowing the cartridge to function in a true reversed-phase manner.

Initially 0.1M solutions of cations in 50% methanol were used to condition the cartridges. These conditions had previously been found to be successful for elution of bases from reversed-phase cartridges [4]. Typical elution profiles for propranolol and atenolol on C2 cartridges are shown in Figure 2. In contrast to cartridges conditioned only with methanol and water (Table 1), conditioning with a salt solution allowed elution of either drug with aqueous methanol or pure methanol. The ability of the cations to block the silanols and so facilitate elution under reversed-phase conditions appears, as expected, to be related to their cation elution strength. The ordering of the cation strength was similar to that previously established during elution from reversed-phase cartridges [4] and to the ordering of cations in classical cation exchange chromatography [7]. In general the silanol blocking ability decreased across the series $TBA^+>Pb^{2+}>TEA^+>K^+>Na^+>Li^+>NH_4^+>H^+$. There were big differences between the extremes, e.g. TBA^+ and H^+, where the latter gave minimal elution and was comparable to a control where the cationic conditioning stage was omitted. In contrast to the use of cations for the elution of basic compounds [4], the differences between some of the ions used although significant were small, and the picture was made more confusing by the non-parallel nature of the elution curves (Figure 2a). Surprisingly the ease of elution from C2 and C18 cartridge following treatment with cations was similar.

Although the expected polarity difference between the two cartridge types was not observed, the influence of solute polarity was as expected in that the more polar atenolol eluted at lower eluent strength than propranolol (Figure 2). Apart from a narrower range in elution strength between the ions compared to that seen previously [4], the major difference was the change in the relative position of NH_4^+. Under the present conditions it appeared weaker than the three alkaline earth metals whereas theory [7] and previous experience [4] indicated it to be stronger.

A further interesting feature of this data was the relatively low recoveries, particularly for propranolol (irrespective of cartridge), in the first methanol wash even when strong ions were used as the blocking agent. This suggested that blocking of the ionised silanols was not fully effective under the conditions used, resulting in residual electrostatic interaction.

When used as an eluting agent, 1ml of a 0.1M solution of cation was highly effective at eluting either propranolol or atenolol. Under the present conditions it was not. One

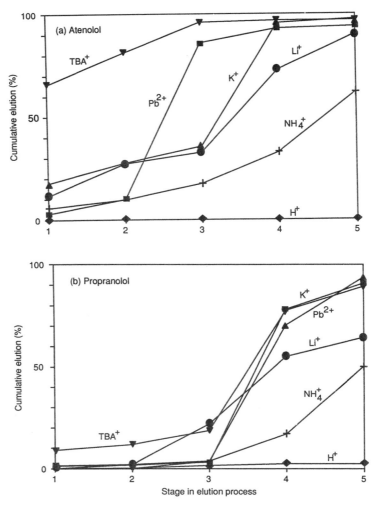

Figure 2.   Cumulative elution profiles for atenolol(a) and propranolol(b) eluted from a C2 cartridge which had been preconditioned with a range of cations. Stage 1 = application, 2 = water wash, 3 = 50% methanol wash, 4 = 1st methanol wash and 5 = 2nd methanol wash.

possible explanation is the fact that when used as an eluting agent the cations probably interact with the limited number of highly reactive silanols to which the drug is bound. When used as conditioning agent however it is assumed that the cations must interact with all the residual silanols to be fully effective. The mass of solute used in the present experiment was of the order of $0.004\mu mol$ but the number of unbonded silanols was $490\mu mol$ for a 100mg cartridge. The latter figure was calculated, assuming a surface silanol concentration of $7\mu mol/m^2$, a surface area of $600m^2/g$ and 50% coverage of bonded phase. However, a 1ml aliquot of a 0.1M solution of the cationic reagent contains only $100\mu mol$ of M^+. It would appear, therefore, that under the conditions described above there was insufficient cations to block every silanol present, assuming they were all ionised.

A further experiment was therefore carried out using three concentrations (0.01, 0.1 and 1.0M) of two cations (Na^+ and NH_4^+) in the conditioning solution. The data for ateno-

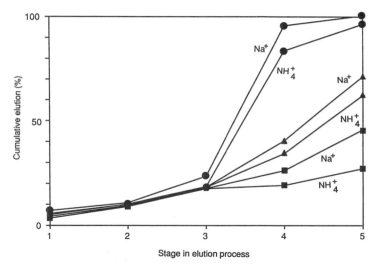

Figure 3. Cumulative elution profiles for atentolol eluted from a C2 cartridge which has been preconditioned with Na^+ or NH_4^+ in 50% methanol at 3 different concentrations, 0.01M (■), 0.1M (▲) and 1.0M (●). Elution stages as Figure 2.

lol on a C2 cartridge (Figure 3) shows more facile elution with increasing concentration of both NH_4^+ and Na^+. This supports the view that a cation mass of >490µmol (i.e. 1ml of 1M solution) is necessary to block the ionised silanols on the cartridge and allow the stationary phase to act in a purely reversed-phase manner.

At this point it was considered of interest to investigate the effect of the cation conditioning solvent. In this work two cations (Na^+ and K^+) were prepared at the sub-optimal concentration of 0.1M in either methanol, 50% methanol or water and the cartridges condi-

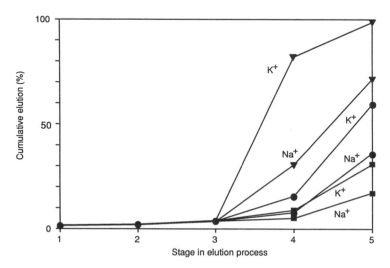

Figure 4. Cumulative elution profiles for propranolol eluted from a C2 cartridge which has been preconditioned with Na^+ or K^+ (0.1M) in either methanol (▼), 50% methanol (●) or water (■). Elution stages as Figure 2.

tioned as described above. The results of this experiment with propranolol as the test compound are shown in Figure 4. It can be clearly seen that the nature of the conditioning solvent has a profound effect on the ability of the cations to block the silanol groups. To allow the stationary phase to act in a purely reversed-phase manner, it would appear that the silanol blocking agent (K^+ or Na^+ in this instance) should be applied in methanol rather than water as used by most workers in the field [5,8-10]. Although it is unclear why this should be so it may be related to the way the cations are solvated in the different media. In contrast to the situation in aqueous solution, in methanol, the lipophilic solvation sphere may allow greater penetration of the solvated cation through the lipophilic hydrocarbon layer and hence greater access to the silanols.

To convert a mixed mode reversed-phase cartridge into a cartridge exhibiting a purely reversed-phase mechanism, conditioning procedures radically different to those reported in the literature must be used [e.g. 5,8-11]. Although the reasons for conditioning with buffer are rarely stated, the conditioning steps used are obviously non-optimal in terms of blocking silanols and probably represents a poorly controlled variable in a number of analytical procedures.

CONCLUSIONS

Experiments have shown that effective blocking of the silanols on reversed-phase extraction cartridges by preconditioning with cations is possible. The necessary conditions however, involving high concentrations of salt solutions (1M) and methanol as solvent rather than water, are different to those frequently employed in solid-phase extraction procedures. Apart from the narrow range and low strength observed for the ammonium ion, the ordering of the cation strength was similar to that defined previously involving cation-exchange processes.

With a few exceptions [3,9], little attention or justification has been given to the conditioning stage of solid-phase extraction methods. It would appear that many workers merely treat the cartridge with the same buffer as used to dilute or buffer the sample prior to application. These data suggest that more consideration should be given to this stage in the extraction procedure. In particular, full details of all buffers used should be presented. Given the difference in strength between sodium and potassium cations - common buffer constituents - it is less than satisfactory to merely describe a conditioning reagent as 'phosphate buffer'.

ACKNOWLEDGEMENT

I wish to thank Nicola Ward for her practical assistance.

REFERENCES

1. P.M. Harrison, A.M Tonkin, C.M. Cahill and A.J. McLean. The rapid and simultaneous extraction of propranolol, its basic metabolites and conjugates from plasma and assay by high performance liquid chromatography. *J. Chromatogr.* 343, 349-358 (1985).
2. G. Musch, Y. Buelens and D.L. Massart. A strategy for the determination of beta blockers in plasma using solid-phase extraction in combination with high-performance liquid chromatography. *J. Pharm. Biomed. Anal.* 7, 483-497 (1989).
3. G. Musch and D.L. Massart. Isolation of basic drugs from plasma using solid-phase extraction with a cyanopropyl-bonded phase. *J. Chromatogr.* 432, 209-222 (1988).
4. B. Law, S. Weir and N.A. Ward. Fundamental studies in reversed-phase liquid-solid extraction of basic drugs I: ionic interactions. *J. Pharm. Biomed. Anal.* 10, 167-179 (1992).
5. R.J. Ruane and I.D. Wilson. The use of C18 bonded silica in the solid-phase extraction of basic drugs - possible role for ionic interactions with residual silanols. *J. Pharm. Biomed. Anal.* 5, 723-727 (1987).

6. R. Bland. Application of solid-phase extraction to the determination of salbutamol in human plasma by HPLC **in** Third Annual International Symposium on Sample Preparation and Isolation using Bonded Silicas, Analytichem International, Harbor City, CA, 1986, p93.

7. J.C. Kraak. **In** C.F. Simpson (ed.) Techniques in Liquid Chromatography, Wiley, Chichester (1982), p313.

8. V. Marko, L. Soltes and I. Novak. Selective solid-phase extraction of basic drugs by C18-silica. Discussion of possible interactions. *J. Pharm. Biomed. Anal.* 8, 297-301 (1990).

9. R. Kupferschmidt and R. Schmidt. Highly specific 'off-line' solid phase extraction of basic lipophilic drugs from blood plasma with automated 'on-line' elution. In: Eleventh International Symposium on Column Liquid Chromatography Abstracts of Papers, Amsterdam, 1987, p124.

10. H.T. Karnes, K. Opong-Mensah, D. Farthing and L.A. Beightol. Automated solid-phase extraction and high-performance liquid chromatographic determination of ranitidine from urine, plasma and pertoneal dialyzate. *J. Chromatogr.* 422, 165-173 (1987).

11. R. Walker and I. Kanfer. High-performance liquid chromatographic analysis for cyclizine and its metabolites, norcyclizine in biological fluid using solid-phase extraction. *Chromatographia* 24, 287-290 (1987).

EXPERIENCES WITH AUTOMATED SAMPLE PREPARATION IN BIOANALYSIS

V.S. Picot and R.D. McDowall

Department of Bioanalytical Sciences
Wellcome Research Laboratories
Beckenham
Kent, BR3 3BS. UK

SUMMARY

The automation of sample preparation within bioanalysis has been largely ignored and may now be regarded as the weak element in automated analysis. The reason is that sample preparation can be applicable to one method but not another. In addition, there are factors such as variation in the available sample volumes and the consistency or viscosity of biological samples, dilution or concentration steps may be required for some, but not all, samples. The particulate or semi-solid nature of some matrices may adversely affect an automated analysis if not taken into account.

Sample preparation is often considered as one of the rate limiting factors in many analytical methods. However it is probably the most critical factor for determining the accuracy and precision of analytical results, as well as for sample throughput and turnaround time. For these reasons, automation of sample preparation is now receiving more attention than before and more automatic systems are now being manufactured.

Here we discuss our experience with a wide range of automated sample preparation techniques, within the context of analysis for compounds (both endogenous and exogenous) in biological fluids.

LABORATORY AUTOMATION AND SAMPLE PREPARATION

Laboratory automation can be classified into two areas: laboratory management automation and instrument automation [1]. The former area is concerned with management of the data and information produced by the laboratory, typified by laboratory information management systems [2,3], whilst the latter area covers increased sample throughput with more efficient data capture and processing. Instrument automation can be classified further into two main types: flexible and dedicated [4,5]. However, the integration of both instrument and laboratory management automation is necessary to achieve overall laboratory objectives such as:

Sample Preparation for Biomedical and Environmental Analysis,
Edited by D. Stevenson and I.D. Wilson, Plenum Press, New York, 1994

- Greater increases in productivity and quality of results (either in numbers of samples assayed per unit time or speedier turnaround time).
- Systems that should be capable of the same or better precision and accuracy as the existing manual methods.
- The freeing of trained laboratory staff to do more productive work than extracting samples
- Reduced human contact with biohazards.
- Release of staff from undertaking tedious tasks thus improving morale.
- Cost savings.
- Extended or unattended operations.

The main thrust of laboratory automation should be towards creating strategic advantage for an organisation [5].

Instrument automation has been available, in various forms such as autosamplers, autoanalysers, integrators etc for many years. However the main emphasis of this area of automation has been towards instrumental analysis and data acquisition and not towards sample preparation.

The automation of sample preparation within analysis has been largely ignored and may now be regarded as the weak element in automated analysis (with some notable exceptions such as clinical chemistry analysers). The reason is that sample preparation can be highly application specific: applicable to one method but not another. There are other factors such as variation in the consistency or viscosity of biological samples. Dilution or concentration steps may be required for some but not all samples and the particulate or semi-solid nature of some matrices may adversely affect an automated analysis.

Sample preparation is often considered as one of the rate limiting factors in many analytical methods, although it is probably the most critical factor for determining the accuracy and precision of analytical results as well as for sample throughput and turnaround time [5]. For these reasons, automation of sample preparation is now receiving more attention than before [6-8] and more automatic systems are being manufactured or conceived [9].

Here we discuss our experience with a wide range of automated sample preparation techniques, within the context of analysis for compounds (both endogenous or exogenous) in biological fluids. The automated analysis of tissue samples is not considered, unless suitable pretreatment has been used to obtain a liquid extract which can then be analyzed automatically using the principles outlined in this paper.

CHOICES FOR AUTOMATED SAMPLE PREPARATION

As indicated above, instrument automation can be sub-divided into flexible and dedicated automation:

Flexible Automation

Flexible automation can be reprogrammed and also re-engineered to change the task that can be undertaken by it. This type of automation is typified by robotic arms. A description of the robotic arms and operations they can undertake is beyond the scope of this paper and the reader is referred to the reviews by Dessy [10,11] and Last [12]. Moreover, general applications of robotics and specific sample preparation examples can be found in the series 'Advances in Laboratory Automation - Robotics' [13-18]

The acquisition of a robotic system can take the analyst into an area in which he has little experience: often the task must be specified or designed in comparison to the purchase and use of equipment off-the-shelf [19]. The main advantage of robotic systems lies with their flexibility, as they can be modified or re-engineered to undertake new tasks. Obviously, a modification of an existing procedure is easy to achieve and this can be done relatively quickly depending on the degree of work involved. If the task is completely different, then it can be more involved, especially if the work requires the custom manufacture of specific peripheral components. In our experience, input from engineers is very helpful and shows the multidisciplinary nature of robotics. Once installed and validated, an application is relatively simple to run and is reliable in operation.

Dedicated Automation

Dedicated automation units are usually limited to a specific task, which may be varied by programming. This can take the form of autosamplers, sample processors, continuous flow analysers and high performance liquid chromatoraphy (HPLC) column switching systems. All offer a cost effective means of automating an instrumental assay. Here, discussion of dedicated systems will be limited to sample processors and HPLC column switching systems (the ASTED is discussed by Buick and Fook Sheung in this volume).

Choosing Flexible Or Dedicated Automation

A way of assessing a laboratory's needs for automation is to break down the analytical method into the unit operations [20]. Once this has been achieved, critically assess what is required; if liquids are moved or transferred from one location to another then a dedicated instrument may be the best approach. If there is extensive manipulation of the sample e.g. homogenization of a biological tissue, then the manipulative advantages of a robotic arm should be apparent.

Comparison Of Automated Systems

A comparision of a manual and three automated techniques (robot, Gilson-AASP and column switching) used to analyse compounds in plasma has been presented [21]; all are sufficiently precise and accurate. However the type, the degree of automation and the costs that a laboratory can pay are varied. Flexible robotic systems are expensive and usually the most complex to implement, but can be justified with a high throughput assay and can also undertake manipulative tasks that the dedicated systems cannot. Dedicated automation is more cost-effective and quicker to implement, but has the disadvantage of being limited to one task. However, dedicated automation has the potential to provide the greatest throughput with the lowest cost for an automated system.

METHOD DEVELOPMENT AND ESTABLISHMENT: WHEN TO AUTOMATE?

The analyst is faced with a problem when starting to develop or establish a method: should the assay be automated immediately or be developed manually first and automated later? While no concrete rules can be developed, it is our experience and opinion that it is easier for methods to be developed manually first, then automated later. The main exception to this statement is column switching, this is an automated technique and can only be developed in this manner; suitable guidelines are given by Huber and Zech [22].

Within an industrial context, time constraints are the main reason for this approach. A method may be required relatively rapidly and it is usually quicker to develop a manual method than an automated one. This allows control to be exercised over the method, it's principles and operation can be understood, and trouble shooting problems becomes easier without the added complication of automated equipment. This approach has the added benefit that results can be generated more rapidly, allowing work to proceed whilst a method is refined and automated.

However, if an effective automated method has already been developed for a compound with similar extraction and detection properties to the analyte in question, it can be modified for use with new compounds. Thus these analytes can be analyzed automatically from an early stage.

Attempts to develop automated methods from first principles require that sufficient time, human skills and resource are available and are not critical elements in the method development equation. It is important that the method selected comprises sample preparation processes which are amenable to automation. The two major techniques that have the greatest potential for automation are solid phase extraction and HPLC itself. Other methods such as protein precipitation and ultrafiltration are difficult to automate or may not be easily justified on the basis of a cost/benefit analysis [8]. It is possible to automate liquid-liquid extraction using robotics or continuous flow methods. However, in our view these will not be financially viable unless manipulation of the sample is required, in the light of recent developments in sample processors.

CONSIDERATIONS FOR THE AUTOMATION OF SAMPLE PREPARATION

In order to use automation effectively, the following factors should be considered before attempting any work on a method:

Analytical Objectives And Assay Complexity

The following questions should be answered: how many analytes are to be measured now? Will this change over time or in the light of any knowledge of the compound's metabolism? What is the concentration range over which each analyte is to be measured? How many animal species, and how many matrices, will the method need to be validated in? What limits of quantification are required and from what sample volume? Is the analyte bound to plasma protein? The answers to these questions will give the scope and limitations of the assay procedure. It is relatively easy to design an automated assay when these questions have been answered first. To modify or tinker with an existing method several times is inefficient. This overall approach should also give a better understanding of the total resource required for the work.

Total Sample Numbers

The total number of samples to be analysed can be defined either per year or over the life time of a project. In our opinion there is little point developing an automated system for less than 250 - 500 samples as manual methods can usually be developed and implemented more rapidly. However, when an already established method is available for a structurally related compound, a slight modification may enable an automated method to be developed in minimal time.

Rate Of Sample Arrival

How will the samples arrive and what is the turnaround time required? While the overall numbers per annum may indicate that automation is required, the actual numbers arriving in the laboratory and the turnaround time may dictate otherwise e.g. in a laboratory dealing with emergency poisoning cases, samples may arrive at a continuous rate of a few samples per day, but a rapid turn around is required. It may then be appropriate to develop an assay that has a simple manual sample preparation and chromatographic separation appropriate to the objectives of the analysis. Conversely, if samples arrive in larger numbers and turnaround time is not as vital, an assay can be set up and run when sufficient samples have accrued to make an automated analysis cost-effective.

Batch Size

The number of samples per batch should be considered in relation to the analytical objectives. A large number of batches containing many samples would benefit most from automated analysis as this allows the widest choice for an automated system: robotics, sample processors or column switching systems.

Analyte Stability

Poor analyte stability can have a dramatic effect on the type of automated system used or in the most extreme example can dictate whether or not automated analysis can be used at all. If a compound is temperature labile, then the use of cooled sample racks or trays may minimise decomposition; if degraded enzymatically an inhibitor may need to be added to prevent breakdown of the analyte pending analysis. Analyte instability can have a profound influence on the number of samples processed per unit time.

Number Of Analytes

The number of analytes and their polarity range are factors influencing the choice of both the method of extraction and the whole automated system.

Matrix Composition And Effects

Every biological matrix has a different chemical composition which can affect the performance of an assay. For example, plasma has a high proportion of protein compared with urine and is usually centrifuged to remove precipitated fibrin before it is processed. Urine has a larger proportion of polar, low molecular weight compounds, so samples may require dilution to reduce salt effects. These differences can affect how an automated analysis is set up and operated.

Matrix Consistency

The consistency of a biological matrix can vary from a liquid to a solid. However, most instrumental techniques require that a sample is liquid for analysis. Therefore, unless the automation includes a homogenisation step, most automated systems are usually restricted to liquid samples only. To ensure a reliable assay when analyzing plasma or serum samples using many sample processors, we recommend that each sample be diluted with an appropriate solvent prior to processing. This serves to both reduce the viscosity of the sample and help disrupt weak protein binding effects.

Sample Preparation Technique

The techniques used for preparing the sample for analysis must be amenable to automation. Therefore, to automate an assay, extraction methods using liquid-solid extraction (solid-phase extraction, SPE) or the HPLC column itself should be used, as it is relatively easy to automate the separation of a liquid and a solid compared with two liquid phases.

Time

The time available can often determine the approaches taken in method development. Such constraints may not always be compatible with the development of automated methods.

Limit Of Quantification (LOQ)

The lower the LOQ required for the analysis, the longer the method may take to develop. This is true whether an automated or a manual method is being developed.

Assay Complexity

The processes involved in an automated assay should be as simple as possible. Complex automated assays requiring sophisticated on-line sample preparation and post column reaction detection will take a long time to develop into routine and robust methods.

Human Skills

Never forget the human resource that is available to develop an assay. Do you have the right people to develop the assays you require?

AUTOMATION OF EXISTING ASSAYS

Automation of an existing assay can be achieved using either robotic arms or dedicated autosamplers. If an existing assay is to be automated by autosamplers it may need to undergo some modifications whilst in our view, a robotic system is best used to automate and reproduce specific human activities.

Certain sample preparation techniques are more amenable to automation than others; the easiest to automate are SPE and dialysis. SPE can be automated either by dedicated autosamplers or using a liquid chromatograph via column switching. Some robotic systems and autosamplers can be used to automate liquid-liquid extraction however, automation of this procedure may be inefficient compared with a manual method and may not be financially viable when automated LC or column switching is available.

Once an assay has been automated, the analyst is free to undertake other work in time that would otherwise be spent in routine manual extractions. However if SPE or column switching are selected as the means of automating sample pretreatment, the early stages of the analysis e.g. such as centrifugation and possibly dilution of the samples may still need to be done manually.

DIRECT INJECTION OF MATRIX: HPLC COLUMN SWITCHING

In the case of the direct injection of the matrix the first stage (removal of solid material) still needs to be done manually i.e. centrifugation at 12,000 rpm for two minutes. If filtration via membranes is used there is the potential problem with small sample volumes and adsorption of analytes to the membrane.

Measurements of analytes in plasma by direct injection have potential problems associated with plasma proteins. Injection of particulate matter can block the injection system or column. The use of wide bore tubing (size 0.5 mm id) can help minimise this together with dilution of the sample. Blocking of the injector port is a problem with the Gilson-AASP, a hybrid between automated SPE on individual columns and column switching. The system uses low pressure rated syringes and tubing to hold the sample train and pump it through the AASP cartridge. Centrifugation of the sample is essential, followed by dilution with a large volume of aqueous buffer.

Adjustment of pH plays an important role in optimising the extraction of an analyte from plasma onto an extraction cartridge or HPLC column. In the analysis of zaprinast from plasma an acidic medium was required for optimal recovery with minimal interference. Acetic acid was used for pH adjustment, which did not result in any precipitation of plasma proteins, and only small volumes of plasma (20µl) were used for analysis [23]. If on line sample extraction is performed from acid medium resulting in protein precipitation, the acid and plasma must be thoroughly mixed and centrifuged before automated analysis can begin with the supernatant.

Column switching has the potential for trace enrichment by loading large volumes of samples, up to ≈200µl of plasma, onto an extraction column. The large volume makes it especially important to remove solid or precipitated material. The extraction column is frequently packed in the laboratory and is prepared from large particle material (20-40µm) to help prevent blocking. However, if a largely aqueous mobile phase is used (< 5% organic) and no solid material is present in the sample, commercially available pre-packed 5µm cartridge columns can be used satisfactorily. Even if replaced daily, or after circa 100 samples, these columns are financially a viable alternative to solid phase cartridges. The main problem with this approach is the supply of sufficient numbers from the distributors. In our experience, such columns may last considerably longer than 100 injections as long as a suitable clean up stage is employed after each injection. This may constitute using a gradient with up to 30% organic modifier, so that most compounds retained on the cartridge are removed. This will increase the analysis time per sample so that run times less than 25 minutes per sample are not easily achieved. Thus the total number of samples processed per day is not normally increased by column switching but manual time for sample preparation is considerably reduced.

The choice of extraction phase used in SPE or column switching is important. The phase should be less retentive than that of the analytical column. Analytes can be concentrated on the top of the analytical column in heart-cut mode or back-flushed with a small volume of the final mobile phase.

Success with direct injection can also be achieved using hydrophobic PLRP polymer columns with large (100 A) pore sizes. These were initially developed for protein separation, as the large pores enable proteins to pass through unretained, hence they are useful for direct injection of plasma. However, peak shape may not be as good as that obtained from conventional chemically modified silica phases.

Direct injection of plasma and urine can only be performed into polar highly aqueous solvents. Precipitation of proteins with acetonitrile, followed by centrifugation, means that a higher proportion of organic modifier can be used in with the mobile phase. If this technique is used, care must be taken to see that the analytes are not lost with the precipitate. Injection of samples containing a relatively high percentage of organic solvent into aqueous mobile phases may also result in band broadening which can lead to loss of column resolution and efficiency and ultimately lower assay sensitivity.

Small hydrophillic compounds may be difficult to extract using either SPE cartridges or reversed-phase packings, due to poor retention and or the inability to provide sufficient sample clean-up. Removal of interferences may often be accompanied by the removal of the analyte! The use of a long extraction column (>5cm) in a column switching system allows separation of the analyte away from endogenous material which passes to waste. The analyte can then be heart cut onto an analytical column, a more controllable approach. For this to work effectively, other external factors must be monitored closely, e.g. column temperature must be kept constant. The extraction column must be allowed to equilibrate fully after clean up before the next injection once the work has been performed, otherwise a decrease in retention may result. Over time, an analyte can drift outside of the preset times used to switch valves for heart cutting; this will result in a decrease in peak size, which may not necessarily be reflected in changes in the peak size of any internal standard.

Experiences Of Analyte Extraction

Ion exchange phases may retain large amounts of endogenous material. Hydrophobic weak acids and bases are easy to work with and good methods easily produced. Strong bases interact with residual silanol groups, these interactions can be utilised for selective isolation, a fact not realised in the early use of SPE.

An additional problem for on-line sample preparation using SPE is that once retained, small polar molecules may be removed by the mobile phase in a broad band, not in a small volume, if the mobile phase is highly aqueous or the pH has not been optimised for elution. Therefore this can present a sensitivity problem if there is no concentration step following elution. The Gilson-AASP combination can overcome this problem by on-line elution from the extraction cartridge. The cartridge used should have a bonded phase less hydrophobic than that of the analytical column, so that mobile phase will elute the analyte from the cartridge. Elution with a solvent of increased organic strength results in a smaller volume but poses a new problem that its injection into a highly aqueous mobile phase frequently results in band broadening. This problem of concentration is being addressed by some automatic systems that have automated solvent evaporation capability (e.g. Millilab and Benchmate).

CONCLUSIONS

Automation, in various forms, can offer the bioanalyst the means of improving laboratory efficiency. Before implementing a system, a method should be carefully evaluated in relation to factors such as the analytical objectives, the resources available and the time required to develop the assay before starting the work. Consideration of the physico-chemical properties of the analyte, the biological matrix and the silica bonded phase used for isolation will aid the efficient development of robust methods.

REFERENCES

1. J.G.Liscouski. Laboratory Automation for Computers in the Laboratory, Current Practice and Future Trends, Ed J.G.Liscouski, American Chemical Society Symposium Series 265, American Chemical Society, Washington DC, 1 (1984).
2. R.R.Mahaffey. LIMS Applied Information Technology for the Laboratory, Van Nostrand Reinhold, New York, 1990.
3. R.D.McDowall Ed. Laboratory Information Management Systems: Concepts, Integration and Implementation, Sigma Press, Wilmslow, Cheshire, 1988.
4. R.W.Arndt. Automating the status quo. *Chem. Brit.*, 22, 974 (1986).
5. M.Linder. Laboratory automation and robotics - quo vardis? In: Scientific Computing and Automation (Europe), ed E.J.Karjalainen, Elsevier, Amsterdam, 1990, 273 (1990).
6. R.D.McDowall, J.C.Pearce and G.S.Murkitt. Sample preparation using bonded silicas: recent developments and new instrumentation. *Trend. Analyt. Chem.* 8, 134-140 (1989).
7. H.G.Fouda. Robotics in biomedical chromatography and electrophoresis. *J. Chromatogr.* 492, 85-108 (1989).
8. D.C.Turnell and J.D.H.Cooper. Automation of liquid chromatographic techniques for biomedical analysis. *J. Chromatogr.*, 492, 59-83 (1989).
9. H.M.Kingston. Consortium on automated analytical laboratory systems. *Analyt. Chem.* 61, 1382A-1384A (1989).
10. R.E.Dessy. Robots in the laboratory, Part 1. *Analyt. Chem.*, 55, 1100A-1114A (1983).
11. R.E.Dessy. Robots in the laboratory, Part 2. *Analyt. Chem.*, 55, 1232A-1242A (1983).
12. P.E.Last. The use of robotics in the pharmaceutical industry. *Chem.Brit.*, 23, 1073-1076 (1987).

13. J.R.Strimaitis and G.L.Hawk (Editors) in Advances in Laboratory Automation - Robotics Vol 1, Zymark Corporation, Hopkinton, MA, USA (1984).
14. J.R.Strimaitis and G.L.Hawk (Editors) in Advances in Laboratory Automation - Robotics Vol 2, Zymark Corporation, Hopkinton, MA, USA (1985).
15. J.R.Strimaitis and G.L.Hawk (Editors) in Advances in Laboratory Automation - Robotics Vol 3, Zymark Corporation, Hopkinton, MA, USA (1986).
16. J.R.Strimaitis and G.L.Hawk (Editors) in Advances in Laboratory Automation - Robotics Vol 4, Zymark Corporation, Hopkinton, MA, USA (1987).
17. J.R.Strimaitis and G.L.Hawk (Editors) in Advances in Laboratory Automation - Robotics Vol 5, Zymark Corporation, Hopkinton, MA, USA (1988).
18. J.R.Strimaitis and G.L.Hawk (Editors) in Advances in Laboratory Automation - Robotics Vol 6, Zymark Corporation, Hopkinton, MA, USA (1989).
19. J.C.Pearce, M.P.Allen, S.A.O'Connor and R.D.McDowall. Automation using robotics in the analysis of SKAF 94836 in plasma. *Chemometrics Intell. Lab. Sys.*, 3, 315-319 (1988).
20. R.D.McDowall, E.Doyle, G.S.Murkitt and V.S.Picot. Sample prearation for the HPLC analysis of drugs in biological fluids. *J. Pharm. Biomed. Anal.* 7, 1087-1096 (1989).
21. R.D.McDowall. Sample preparation for biomedial analysis. *J. Chromatogr.* 492, 3-58 (1989).
22. R.Huber and K.Zech, in R.W.Frei and K.Zech (Editors) Selective Sample Handling and Detection in High Performance Liquid Chromatography Part A, Elsevier, Amsterdam, p81 (1988).
23. V.S.Picot, E.Doyle and J.C.Pearce. Analysis of Zaprinast in rat and human plasma by automated solid-phase extraction and reversed-phased high-performance liquid chromatography. *J.Chromatogr.* 527, 454-458 (1990).

CLINICAL ANALYTES FROM BIOLOGICAL MATRICES

Ian D. Watson

Department of Clinical Biochemistry
Aintree Hospitals
Fazakerley Hospital
Longmoor Lane
Liverpool, L9 7AL

SUMMARY

Sample preparation for clinical analysis encompasses a wide range of potential analytes and is challenged by a diversity of matrices; consequently there are many solutions.

Sample preparation can be divided into four phases - sample collection, isolation from the matrix, purification and preparation for analysis. The matrices most frequently encountered in clinical analysis are plasma (or serum), whole blood and urine and many will be familiar with the sample preparation challenges that these pose. However it is occasionally required that analytes be isolated from tissue, faeces, saliva, cerebrospinal fluid, sputum, drainage fluids, etc. Each of these matrices presents a peculiar challenge of its own to the analyst.

The isolation of clinical analytes from matrices of biomedical interest is discussed with illustration of how and why this may be done and the possible pitfalls.

INTRODUCTION

The complexity of the matrices encountered in clinical studies hampers analysis in biological samples and often requires some preparation to provide specificity. Sample preparation can be considered in four phases.

i)   *Sample collection*: Procedures are adopted to ensure the integrity of the determinant during collection, transport and storage;

ii)   *Isolation from matrix*: Biological matrices are complex mixtures of compounds and usually have a high protein content. The determinant may be present in low concentrations and bound to protein. The analyst must isolate the determinant from the matrix, usually by protein precipitation and/or extraction;

Sample Preparation for Biomedical and Environmental Analysis,
Edited by D. Stevenson and I.D. Wilson, Plenum Press, New York, 1994

iii) *Purification*: The determinant(s) of interest must be isolated from other compounds likely to interfere, this may be concomitant with (ii) above but often further isolation is required;

iv) *Preparation for analysis*: Preparation may consist of concentration of an extract to achieve appropriate sensitivity, it may be that derivatisation is necessary prior to the analytical step, e.g. G.C.

The peculiarities of sample preparation for clinical analytes are most marked in the first two categories and it is these areas that this article will concentrate upon.

SAMPLE TYPES AND SAMPLE COLLECTION

Serum, Plasma, Whole Blood

The commonest of biological samples is serum (or plasma), whole blood is occasionally used; when anticoagulation is used there is a potential for analytical interference, the containers or their stoppers may cause contamination but these problems are well recognised. Common errors in blood collection include sample haemolysis, causing analyte dilution/interference in a serum assay and excessive stasis on venepuncture increasing the concentration of protein bound analytes, e.g. total serum calcium. Low volume samples requiring transport over long distance, can be collected on cards, e.g. Guthrie cards for screening for phenylketonuria.

Urine

In the case of urine preservatives may be required to stabilise the analyte particularly if it is susceptible to bacterial degradation. The collection of accurate, timed, urine samples (e.g. 24 hours) is a perennial nightmare; so much so that it has been recommended that creatinine clearance is better predicted from serum creatinine than by measurement [1].

Cerebrospinal Fluid

Cerebrospinal fluid (CSF) is analysed in certain circumstances but the need for lumbar puncture requires there to be a good medical reason for such a manoeuvre. CSF has the advantage of being virtually protein-free, which for analytes for which there is sufficient sensitivity would permit direct sample injection onto a high performance liquid chromatograph (HPLC).

Saliva

Saliva measurements are sometimes made as an alternative to the invasive technique of venepuncture. In my experience adults find the process aesthetically distasteful but children will co-operate. Saliva hydrogen ion concentration changes on stimulation of flow [2] and is of variable composition depending on source; parotid and submandibular saliva differ [2]. The change in hydrogen ion concentration will affect ionised compounds; the more lipophilic a compound the better its transfer to saliva. It has been suggested that prediction of plasma drug concentrations from salivary data is best for drugs that are mainly un-ionised at physiological plasma [3].

Stimulation is effected by mastication or gustation; for the former a washed rubber band is preferred to paraffin, wax or parafilm as they will partition lipophilic compounds [4], while for the latter a sapid stimulus, such as citric acid, is preferred [5].

Faeces

Faecal analysis is disagreeable but sometimes necessary. Collection of timed samples is difficult as the mixing of gut contents makes it almost impossible to identify which stools were formed during the collection period or those which might contain the determinant (if exogenous). Two approaches are used to produce a fluid sample for analysis: ashing to reduce the bulk followed by dissolving of the temperature stable remnants or more commonly, homogenisation with an appropriate volume of fluid to produce a faecal slurry, removal of remaining solids and subsequent processing. The unpleasant nature of the sample dictates that the early stages of sample preparation are conducted in a fume cupboard.

Sputum

For a variety of reasons (e.g. penetration of an antimicrobial in chronic bronchitis) it may be necessary to sample sputum. There is risk of infection, particularly of tuberculosis (TB) and sputum is another aesthetically displeasing sample. Sputum is often collected on expectoration but will almost certainly be contaminated with saliva and possibly nasopharyngeal secretion with or without pus. The preferred collection route is trans-tracheal aspiration despite the fact that this is an invasive procedure.

Keratinaceous Tissue

Hair and nail tissues are used to detect exposure to poisons, e.g. lead or drugs of abuse over a period of time. There is a particular problem with contamination due to the low concentrations found. Following washing the keratinaceous tissues may be solubilised in alkali. In longitudinal studies the difference in growth rates of hair from different areas of the head should be noted [6] as should the fivefold difference in growth rate between finger and toenails [7].

Tissue

While tissue may be obtained at biopsy and analysed it is more frequently examined as a post operative sample or as a forensic sample; in the former case it is often important to avoid histological fixatives and in the latter putrefaction can cause significant interference in assays.

Joint And Drainage Fluids

Joint fluid is usually obtained by aspiration and drainage fluid by leakage and while the latter may be significantly contaminated with pus, these are usually quite liquid and are treated as for serum.

THE CONTEXT

In our Department we analyse over 230,000 samples per year with a test:request ratio of 9:1. The overwhelming majority of these are analyses on serum or plasma (>98%) of the others urine accounts for 0.9%, faeces 0.65%, "fluids" 0.45% and "others" <0.1%. Few of the non-serum samples require sample preparation. Sample preparation is performed on some of the serum samples. In Clinical Biochemistry over 80% of the workload is represented by only 20 analytes. To cope with the large workloads a high degree of automation is necessary with minimal sample preparation. Early attempts at automation tried to mimic manual stages of protein precipitation, centrifugation, separation, etc. Continuous-flow systems used dialysis to separate analyte from serum proteins. Subsequently chemistries have been developed that operate in the presence of protein, these are wet chemistries; recently dry film technologies have been introduced that separate whole blood into serum and a layered chemistry proceeds prior to measurement.

The assays that require sample preparation, other than simple dilution, account for around 1% of the workload of a clinical biochemistry laboratory. Analytes that may require sample preparation include: steroid hormones, drugs, trace elements, biogenic amines and amino acids. The infrequency of assay of unusual matrices makes it more important to know what to do once presented with the problem.

THE PROBLEMS

These may be succinctly summarised as high protein content, the presence of many potential interfering compounds, either endogenous or exogenous, in often much higher concentrations than the determinant, the need for a liquid sample for further processing and the risk of infection.

Sample Preparation

Clinical sample preparation is determined by the analytical end step to be applied. A wide range of analytical techniques are used including wet chemistries with photometric detection, spectrophotometry, fluorimetry, electrophoresis, immunoassay and chromatography.

Direct Measurement

Introduction of the sample directly into the analytical system used to be the exception and protein had to be removed first. Now many analytes, e.g. urea, calcium, are measured following dilution of the sample to avoid protein interference. This approach works well in wet chemistry systems but has also been used in gas chromatography in the measurement of blood alcohols (although head space analysis allows for much longer column life!). A relatively recent development has been the introduction of micellar liquid chromatography [8] which allows direct injection onto a HPLC system. However, this approach has not lived up to its initial promise.

PROTEIN REMOVAL

Protein Precipitation

The early classical wet chemistry methods of clinical chemistry frequently used protein precipitation to produce a clear 'protein-free' supernatant [9,10] in these procedures an acid or alkali precipitant with a salt was used. Such precipitants rarely exceed 80% recovery, frequently less [11]. Water-miscible organic solvents have been used as precipitants prior to liquid chromatography. Of these acetonitrile is the most popular [12], but again there can be problems of poor recovery and also late eluting peaks [13].

There are other problems, protein precipitation is not complete, some globulins can remain in solution, and react in the analytical system. Furthermore, such procedures by their nature are dilutional and are best utilised for high concentration determinants. The precipitants can react in the analytical system, eluting as a peak in HPLC or interfering in a spectrophotometric assay (See Table 1).

Table 1. Effect of precipitant on the colorimetric determination of paraquat.

Precipitant	Background Absorbance (396 nm)	Recovery (%)
Alcohol	0.15	43
Trichloroacetic acid	0.07	R
Perchloric acid	0.12	R
Sulphosalicylic acid	0.04	90
Tungstic acid	0.02	R
Zinc Hydroxide	0.01	60

R   =   Reacts with colour reagents.

Dialysis

Dialysis is the classic procedure for protein removal from small molecular weight analytes; the automation of this technique in continuous flow analysers revolutionised clinical chemistry enabling the explosion of workload in that area to occur. The combination of this technique with a trace enrichment column is available commercially for linkage to an HPLC (see Buick and Fook Sheung; Cooper and Dale, this volume).

Ultrafiltration

The technique of ultrafiltration may be used to measure total drug in serum if it is only marginally protein bound [14]. In the author's experience ultrafiltration works best for acidic and neutral drugs; basic drugs tend to bind to the membrane.

Liquid-Liquid Extraction

Liquid-liquid extraction is the other classic sample preparation technique and depends on the partitioning of the determinand between two immiscible liquids. The fraction of analyte in the organic phase (O_h) is:

$$O_f = \frac{PV}{PV+1}$$

Where V is the phase volume ratio, "Oc/Ac" and P is the partition co-efficient. Oc is the concentration in the organic phase and Ac is the concentration in the aqueous phase

A pH of more than 2 from the pKa, i.e. less than 1% ionised, maximises recovery (for a compound with a single pKa).

While it is simple to design extraction systems, liquid-liquid extraction is waning in popularity. This is due to many factors such as safety (the risk of explosion, flammability, toxicity and carcinogenicity) and the technique dependent influence on imprecision and recovery. Liquid-liquid extraction is also labour intensive, difficult to automate and there are problems of analyte adsorption on glassware or solvent mediated decomposition. Furthermore, impurities in the solvent can cause interference, emulsion formation increases the complexity and decreases recovery and there are the delays engendered in the often lengthy evaporation stage.

Liquid-Solid Extraction

As is self-evident from the increasing number of published applications liquid-solid (or solid phase) extraction is the technique of the future. The extraction may be considered as a distribution ratio between the liquid and solid phases:

$$D = \frac{Vs}{Ab} \frac{Ar}{As}$$

Where D = co-efficient of distribution,
      Vs = volume of solution
      Vb = resin bed volume
      Ar = amount of analyte in the resin
and   As = amount of analyte in solution

Procedures available are many and varied using all the solid phases, and more, utilised in HPLC. Although more expensive than liquid-liquid assays in terms of consumables they are less labour intensive and are widely adaptable; method development is easy and predictable (usually). Liquid-solid extraction is readily automated and this is a further advantage.

We have reported previously on the improvements in recovery, throughput, etc. when liquid-solid extraction is compared to liquid-liquid extraction [15]. This was particularly marked for tricyclic antidepressant analysis - previously a tedious assay due to the procedures adopted to avoid adsorptive losses. The advantages and disadvantages of liquid-liquid and liquid-solid extraction have been compared [16].

SPECIAL STAGES

Tissue

The classical technique for extraction of an analyte from tissue is homogenisation in a volume of aqueous solvent, usually acidic. Such procedures often result in poor recovery, are labour intensive and require large volumes of extract. A gentler technique is to use a protease such as *Subtilising carlsberg* [17,18] which results in a liquid product readily manipulable for further extraction. The total destruction of the tissue structure results in far higher yields.

Keratinaceous Tissue

Liquefaction of hair and nails may be achieved with sodium hydroxide and subsequent organic solvent extraction, prior washing is required to minimise contamination. Recently an enzyme method using pronase has been described which increases the yield of the determinant [19]. The analysis of hair for trace elements has been reviewed [20].

Sputum

The technique we developed to liquefy sputum employed the addition of 10% 2,3-dithiothreitol to the sputum followed by incubation [21]. The cycle was repeated if liquefaction was incomplete.

As incubation was at 60°C for 45 mins this also served to pasteurise the sample from which there is a high risk of tuberculosis should the subject be infected. This was then followed by a liquid-liquid extraction procedure. N-acetylcysteine is also a satisfactory liquefactant for many analytes [22].

Determination Of Free Analytes

Classically free, i.e. unbound, analytes such as drugs have been determined using equilibrium dialysis. This works well provided the temperature and degree of agitation are kept uniform, it is also a time-consuming laborious technique. There is peripatetic interest in free drugs, ultrafiltration has been used and was available as a commercial system for anti-convulsants. Ultracentrifugation has also been proposed [23,24]. None of these techniques is ideal either requiring a lot of time, expensive consumables or equipment. However, if an ultracentrifuge capable of producing a sufficient volume of protein free plasma water is available a high batch throughput should be achievable.

Other Procedures

Ashing has been used to prepare samples but due to the harsh conditions employed only the most stable analytes can undergo this process. It is used in trace element analysis of tissues.

Head space analysis is performed for anaesthetic gases and ethanol.

Internal Standards

Much controversy surrounds the use of internal standards in chromatography. Some maintain they can cause imprecision in an assay, other that they compensate for losses. Who is right? They both are.

If a procedure can be shown to have a low imprecision between batches and operators then external standardisation is effective. If the imprecision is high, e.g. recoveries not tightly reproducible then the selection of an internal standard with similar physiochemical properties to the determined can compensate; this is with the additional caveats that it be present in roughly equivalent mass to the determinant and elutes with a capacity ratio within 30% of the detertminant [13]. For an excellent consideration of the role of an internal standard see reference 25.

FUTURE PROSPECTS

There is little doubt that liquid-solid extraction will continue to grow in popularity at the expense of liquid-liquid extraction for clinical samples. The prospects of readily adaptable affinity columns is already with us and could expand if appropriate marketing were to be conducted. It is also likely that supercritical fluid extraction will become of importance. The trend to increasing automation will continue with an ever increasing number of 'guaranteed' assays from vendors of extraction hardware. The net result will be cleaner, more sensitive, more rapid assays which will be ideal for both routine and research applications of clinical analytes in biological matrices.

REFERENCES

1. R.B. Payne. Creatinine Clearance: A redundant clinical investigation. *Ann. Clin. Biochem.*, 23, 243-250 (1986).
2. C. Dawes and G.N. Jenkins. The effects of different stimuli on the composition of saliva in mass. *J. Physiol.*, 170, 86-100 (1964).

3. J.C. Mucklow, M.R. Bending, G.C. Kahn and C.T. Dollery. Drug concentration in saliva. *Clin. Pharm. Ther.*, 24, 563-570 (1978).

4. C. Hallstrom, H.M. Lader and S.H. Curry. Diazepam and N-demethyldiazepam concentrations in saliva, plasma and CSF. *Br. J. Clin. Pharmacol.*, 9, 333-339 (1980).

5. K.W. Stephen and C.F. Spiers. Methods of collecting individual components of mixed saliva. The relevance to clinical pharmacology. *Br. J. Clin. Pharmacol.*, 3, 315-319. (1978).

6. P. Manson and S. Zlotkin. Hair analysis - a critical review. *Canadian Medical Association Journal*, 133, 1865-1868 (1985).

7. O. Suzuki, H. Hattori and M. Asaro. Nails as useful materials for detection of methamphetamine or amphetamine abuse. *Forensic Science International*, 24, 9-16 (1984).

8. L.J.C. Love, S. Zibos, J. Noroski and M. Arunyanart. Direct injection of untreated serum using non-ionic and ionic micellar liquid chromatography for determination of drugs. *J. Pharm. Biomed. Anal.*, 3, 511-522 (1985).

9. O. Folin and H.S. Wu. A system of blood analysis. *J. Biol. Chem.*, 38, 81-110 (1919).

10. M. Somogyi. A method for the preparation of blood filtrates for the determination of sugar. *J. Biol. Chem.*, 86, 655-663 (1930).

11. D.B. Campbell **in** Assay of Drugs and other Trace Organic Compounds (E. Reid, ed.) Elsevier, Amsterdam, 1976, pp106.

12. B. Widdop **in** Clarke's Isolation and Identification of Drugs 2nd ed. (A.C. Moffat, J.V. Jackson, M.S. Moss and B. Widdop, eds.) Pharmaceutical Press, London, 1986, pp3.

13. S.J. van der Wal and L.R. Snyder. Precision of 'high performance' liquid chromatographic assays with sample pre-treatment. Error analysis for the Technicon 'FAST-LC' system. *Clin. Chem.*, 27, 1233-1240 (1985).

14. I.D. Watson. Clavulanate potentiated ticarcillin: high performance liquid chromatographic assays for clavulanic acid and ticarcillin isomers in serum and urine. *J. Chromagr.*, 337, 301-309 (1985).

15. S.J. McIntosh, I.D. Watson, P.W.D. Hughes, A.M. Chambers and M.J. Stewart. Off-line automated sample preparation - experience in a clinical drug investigation unit. *J. Pharm. Biomed. Anal.*, 5, 353-360 (1987).

16. R.D. McDowell, J.C. Pearce and G.S. Murkitt. Liquid-solid sample preparation in drug analysis. *J. Pharm. Biomed. Anal.*, 4, 3-21 (1986).

17. M.D. Osselton, I.C. Shaw and H.M. Stevens. Enzymic digestion of liver tissue to release barbiturates, salicyclic acid and other acidic compounds in cases of human poisoning. *Analyst.*, 103, 1160-1164 (1978).

18. M.D. Osselton. The release of basic drugs by the enzymic digestion of tissues in cases of poisoning. *J. Forensic Sci.*, 16, 299-303 (1976).

19. C. Offidori, A. Carnevale and M. Chiarotti. Drugs in hair: A new extraction procedure. *Forensic Science International*, 41, 35-39 (1989).

20. A. Taylor. Usefulness of measurements of trace elements in hair. *Ann. Clin. Biochem.*, 23, 364-370 (1986).

21. S.J. McIntosh, D.J. Platt, I.D. Watson, A.J. Guthrie and M.J. Stewart. Liquid chromatographic assay for trimethoprim in sputum and saliva. *J. Antimicrob. Chemother.*, 11, 195-196 (1983).

22. P.K. Li, J.T. Lee and L.M. Baker. Before assay liquifaction of pulmonary mucous secretions with N-acetyl-L-cysteine. *Clin. Chem.*, 26, 1631-1632 (1980).

23. L. Jack, C. Cunningham, I.D. Watson and M.J. Stewart. Micro-scale ultracentrifugation as an alternative to ultra-filtration for the determination of the unbound fraction of phenytoin in human serum. *Ann. Clin. Biochem.*, 23, 603-607 (1986).

24. K. McKenzie, I.D. Watson and M.J. Stewart. Comparison of micro-scale ultracentrifugation and equilibrium dialysis for the determination of unbound serum theophyline concentrations. *Ann. Clin. Biochem.*, 26, 185-188 (1989).

25. P. Haefellinger. Limits of the internal standard technique in chromatography. *J. Chromatogr.*, 218, 73-78 (1981).

A RATIONAL APPROACH TO THE DEVELOPMENT OF SOLID PHASE

EXTRACTION METHODS FOR DRUGS IN BIOLOGICAL MATRICES

R.J. Simmonds, C. James and S. Wood

Upjohn Laboratories
Crawley
West Sussex, RH10 2NJ. UK

SUMMARY

There are many types of solid phase extraction cartridge, and many different priming, washing and eluting solvents can be used. So many are the possibilities for a new assay that sensible choice of available options may appear impossible. Whilst a method of extraction can usually be developed quickly, especially if the ubiquitous C18 phase is used, a more selective and robust method can usually be achieved if a range of options is properly explored.

An approach that fulfils this is presented. The scheme includes:
1. Identification of extraction requirements;
2. Screening of cartridges to identify effective phases;
3. Optimisation of adsorption, rinsing and elution stages to achieve best selectivity and robustness;
4. Validation for routine application.

The need to consider all phases and achieve a contrast with subsequent chromatography is emphasised. The utility of the approach is illustrated for different types of drugs and metabolites.

INTRODUCTION

Despite the vast range of phases available for solid phase extraction (SPE), and high performance liquid chromatography (HPLC) even a cursory review of published HPLC bioanalytical assays shows that in about 70% of these the matrix is extracted with a reversed-phase (C8 or C18 type cartridge), followed by chromatography on the same type of phase (C18 or C8).

Whilst this is an understandable and often successful approach to method development, analysts may be overlooking the range of different phases available, and which may offer significant advantages when used in novel combinations. At least part of the problem

Sample Preparation for Biomedical and Environmental Analysis,
Edited by D. Stevenson and I.D. Wilson, Plenum Press, New York, 1994

may be that the number of products is so daunting. Choices have to be made from many possibilities of packing material, and size, and even what appear to be equivalent phases from different manufacturers may show different characteristics in practice. In consequence there are so many possibilities of SPE and HPLC that it is difficult to choose an efficient combination, and all too easy to get "bogged down" when developing a new method.

We present here an efficient strategy, or series of guidelines, for developing a method which incorporates a "mode sequence" from SPE to chromatography; that is, where extraction is effected with one type of phase, and HPLC with a completely different stationary phase. Such methods are, in our experience, more rugged and less sensitive to changes manufacturers make to cartridges or columns. They can also be more easily adapted to biofluids from different sources, or to different biofluids altogether. They are potentially more sensitive because of the reduction of interfering peaks, and are faster in routine application as long running peaks are generally absent.

In the following we have concentrated on "plasma" as the biological matrix, but the guidelines apply equally well to the development of assays for other biofluids, or faeces and solid tissues. Homogenisation, solubilisation or pre-extraction will obviously be required for these before cartridge extraction is attempted.

Since we want SPE and HPLC to take place on different "phases", we always develop the HPLC and the SPE together. Final choice of which combination to use in practice is then made from a number of possibilities of SPE and HPLC.

The strategy employed in this process of method development is shown below:

1. Development of at least 2-3 HPLC systems with different stationary phases. Check for retention, efficiency, resolution and selectivity.
2. Assess adsorption/desorption of analyte(s) from a wide range of SPE cartridges.
3. Smaller number of effective cartridges tested with dilute biofluid.
4. Loading, washing and elution steps optimised for 1-3 cartridges.
5. Biofluid extracts analysed on different HPLC systems.
6. Choice of HPLC and SPE combination made. Method optimised and validated.

There are many excellent ways of developing HPLC methods (e.g. refs.1,2) and this will not be discussed in any depth. It is important however, to have at least two methods of chromatography, with different selectivities, available whilst developing the extraction.

We have had good success with the "typical" conditions outlined in Figure 1. By manipulation of acetonitrile and trifluoroacetic acid (TFA) concentrations, or use of equivalent concentrations of methanol or tetrahydrofuran or [3] acceptable chromatographic characteristics can quickly be developed for a wide variety of drugs and metabolites. An indication of the range of compounds is given in Table 1. These acidic mobile phases are effective

Injection

15 min

Figure 1.   A "typical" HPLC system employing a C8 column (250 x 4.6 mm) with a mobile phase of 20% acetonitrile 10.2% TFA at 2 ml/min at 40°C aimed for retention times for compound and internal standard approximately 6 and 8 mins respectively (R=6) on a column with 7 to 12000 theoretical plates.

Table 1. Reversed phase HPLC with TFA, 0.1-0.6%
in the mobile phase.

Drug	Character
443 C81 (Wellcome)	Aliphatic, basic peptide
29C (Wellcome)	Aromatic, basic
589C (Wellcome)	Aromatic, aldehyde, acidic
Atracurium (Wellcome)	Large, basic
Trospectomycin	Basic, aliphatic
Methylprednisolone	Neutral
MPA	Neutral
U-66,858	Aromatic, neutral
U-53,996H	Basic

with other stationary phases, for example, C18, C2, phenyl (PH) cyanopropyl (CN) and even, with the addition of a salt, strong cation exchange (SCX). With experience, and working with two chromatographs, a number of methods with similar performance, but contrasting selectivities can be set up in a matter of days. These types of HPLC system are at least adequate for initial experiments, and sometimes are even suitable for the assay in its final form. One of their advantages is that they are also quite resistant to plasma proteins, and diluted, but unextracted biofluid, (equivalent to 25 to 50 µl) can be directly injected onto the column. With sensitive detection, the resulting chromatograms give an excellent indication of what problems the extraction (and sample work up) have to address. HPLC systems that give adequate efficiency and the best selectivity for the biological matrix are the first, and logical, choice for use in developing the method of extraction.

A very wide range of cartridges, including reversed-phases, polar phases and ion exchangers (from a single manufacturer) should be assessed, without preconceptions as to what will and will not work, with dilute aqueous solutions, buffered or otherwise, according to the scheme shown below.

1. Priming   -   2 x 1 ml acetonitrile
                 2 x 1 ml water
2. Loading   -   1 ml aqueous solution, 1 to 5 µg/ml
3. Wash   -   2 x 1 ml water
4. Elute   -   1 ml 60% acetonitrile, 0.2% TFA
5. Compare eluate and standard solution HPLC peak heights.

Note that we only use 100, or even 50 mg cartridges. It is important to prime these adequately with organic solvent and water. The eluting solution is a "stronger" derivative of the HPLC mobile phase. At this stage we are simply looking for an on/off mechanism; that is, efficient adsorption and desorption. Note that assessment of recovery is simple, and by duplicating some idea of reproducibility can be obtained. The results of a cartridges screen for an Upjohn analgesic and internal standard are given in Table 2. Here C18, C8, Diol and CN phases were chosen to look at the extraction from 0.5 ml of diluted plasma, sequence detailed below.

The results are shown in Table 3. Note that C8 and C18, which gave good results for aqueous solutions, not only fail to extract the analytes from the biofluid, but also gave "dirty" extracts. The diol phase gave good efficiency, and as this provided such a contrast to the

Table 2. Result of cartridge screen for U-53996H and U-54586 (Internal Std.), extracted from water.

Cartridge	% Elution	
	U-53996H	*U-54586*
C18	75	71
C8	49	33
C2	15	8
PH	--	--
Si	32	35
2-OH	84	71
CN	75	71
CBA	19	12
SCX	--	--
SAX	--	--

HPLC mode, (C8) extracts were clean and free of interfering peaks, even with sensitive UV detection at 240 nm. In general 0.5 ml of biofluids is the most we extract without a more complex sample work up. Small volumes are less likely to clog cartridges, and appear to give cleaner extracts. For some compounds no extraction from the biological matrix was obtained even though efficient adsorption from aqueous solution was found with several cartridges. If this is the case, sample preparation will need to be more than simple dilution in water. A plasma protein precipitation step, or adjustment of pH with acid or buffer solution, or even incubation with proteases may be required to counter protein binding or competition from endogenous compounds.

The cartridge washing, or rinsing, stage can be investigated by successively rinsing a loaded cartridge with 2 ml of 10, 20, 30, 40, 50 and 100% methanol. The eluates are collected and analysed. Obviously, the "strongest" wash that leaves analytes bound is chosen. Different organic modifiers can also be tried, with buffered solutions at acid, basic, and neutral pH, and these may give useful differences in selectivity.

The elution step can be optimised by eluting loaded and rinsed cartridges successively with 0.2 ml aliquots of eluting solution, and collecting and analysing each. A typical histogram is shown for a drug and internal standard from a reversed phase cartridge in Fig.2. Here elution of the drug was complete after 0.6 ml, though 0.8 ml was required for good

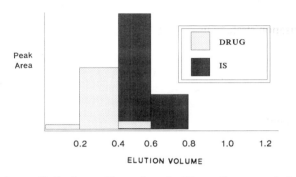

Figure 2. Elution profile for drug and internal standard from a C_{18} reversed-phase cartridge.

Table 3.   Extraction of U-53996H and U-54586 from
            0.5 ml plasma diluted to 1 ml with water.

| Cartridge | % Elution | |
	U-53996H	U-54586
C18	1.0	1.5
C8	3.0	3.0
2-OH	100.0	100.0
CN	88.0	77.0

recovery of the internal standard. Obviously, the smallest volume with good and reproducible recovery was chosen. Again, different organic solvents, or mixtures, or buffered solutions may be tried. We always aim for an eluting volume of no more than 0.6 ml, and 60% acetonitrile is the solvent of choice, since it is simple to partially evaporate by increasing the vacuum in the extraction manifold (for 5 to 10 minutes). The eluate will reduce to 0.2 ml quite repeatedly, and nearly all of this (now aqueous) extract can usually be injected on to the HPLC column.

What perhaps is best emphasised here is that we want all steps in our extraction procedure to be "non-critical" so that small changes in reagent make up, or variations in volumes, do not affect reproducibility. Following the guidelines described, this is generally the case. However, it is a good idea to look at small variations in procedure, and to try different batches of cartridge, reagents or sources of biofluid to demonstrate that this is indeed so. Methods that appear only to work with one batch of cartridges, or where one of the steps is unduly sensitive to concentration or volume of reagent, may point to adsorption that depends upon uncapped silanol residues in the cartridge phase. Such methods are to be avoided. Perusal of the results of the full cartridge screen will generally indicate the mechanism of adsorption and desorption; the results for the silica cartridge will indicate the importance, or otherwise, of uncapped silanol groups. However, we usually avoid these cartridges because they tend to be unreliable for aqueous samples.

At this point in the development of a method we will usually have two or three methods of extraction, and at least two methods of HPLC.

Eluates from different cartridges are now injected onto the HPLC, and the compatibility of extracts with the methods of HPLC, and the cleanliness, or otherwise, of the chromatograms assessed. The best combination(s) can then be chosen and HPLC and extraction optimised together.

The optimisation strategy used is as follows:

1. Compare equivalent columns from different manufacturers;
2. Modifications to mobile phase:
    a) Buffer salts
    b) Ion pair
    c) Solvent triangle
3. Purity/source of solvents and reagents;
4. Preparation of cartridges;
5. Preparation of samples;
6. Change wavelength of detection (UV) or wavelengths (fluorescence);

These are fairly obvious, and do not need to be discussed in detail. Buffered ammonium acetate solutions either as eluent or mobile phase may give cleaner chromatograms than TFA, and the quality of reagents can be all important. "Reagent blanks", where water in place of the biofluid is extracted will give an indication of possible problems. It is good practice to prime cartridges with the eluting solution or similar. This can reduce interfering peaks, for example plasticisers, that derive from cartridges.

It is important that the complete method is then validated. Some of the more important factors that should be looked at with these sorts of methods include the batch of the HPLC column, the batch of the SPE cartridges, sources of solvents and reagents, sources of biological fluids the effect of deliberately "sloppy" sample preparation and the extraction of real samples. The need to look at actual samples from a real study should be emphasised. Always look at possible sources of variability.

Though we have had good success with developing methods using these procedures there are some provisos that should be mentioned:

a)   The internal standard needs to be structural analogue of the compounds to be measured. The extraction can be so selective that readily available analogues may be lost either in the rinsing stage, or are not eluted. Good synthetic back up, where an internal standard can be designed and made is necessary if an internal standard really is essential to the assay [4]. However, given efficient extraction the repeatability of the method can probably be adequate without an internal standard [5];

b)   Modification of an assay, for example to measure metabolites, is difficult (due to the built-in selectivity of the procedure), and it may be better to develop a new extraction, rather than modifying the old one. Compromises will have to be made to ensure that dissimilar analytes (for example the parent drug and metabolites) are efficiently extracted;

c)   A large factor in our strategy is the use of TFA in both mobile phase and the cartridge eluting solution. Choice of the eluting solution is obvious critical to the results of the cartridge survey. Pure organic solvents are generally ineffective as eluting solutions for cartridge screening. Basic compounds, for example, may not be eluted from any of the polar phases, or from those reversed-phases which are not completely endcapped. If TFA is incompatible with the analytes or HPLC, then a derivative of the HPLC mobile phase is used, but modification to increase its eluting "power" is desirable;

Incidentally, the above consideration shows why we prefer "off-line" SPE to "on-line" methods using, for example, the AASP. Restriction of the eluting solution to the mobile phase can limit the choice of cartridge phase, and hence the selectivity obtainable in single stage extraction.

Methods have been developed following the described scheme which include a wide range of compound types and which include acidic, basic and neutral compounds (see also refs.4-7).

Only in the case of the antibiotic trospectomycin was there no contrast between SPE and HPLC phase, though this is a special case [4]. However, there was a contrast between the eluting solution and the HPLC mobile phase.

We hope that the strategy described will encourage others to explore the wide range of SPE cartridges and HPLC columns available. In practice viable methods can be developed very quickly, and have given few problems in routine use, whether in-house or at contract laboratories.

REFERENCES

1. M. De Smet and D.L. Massart. Automated method section for HPLC analysis of drugs. *Trends in Anal. Chem.* 6, 266-271 (1987).
2. P.J. Schoenmakers, H.A.H. Billiet and L. De Galan. Use of gradient elation for rapid selection of iocratic conditions in reversed phase high performance liquid chromatography. *J. Chromatog.* 205, 13-30 (1981).
3. R. Lehrer. High performance LC with four solvents. *Int./Lab.* Nov/Dec, 76-88 (1981).
4. R.J. Simmonds, S.A. Wood and M.J. Ackland. A sensitive high performance liquid chromatography assay for trospectomycin, an aminocyclitol antibiotic, in human plasma and serum. *J. Liquid Chromatog.* 13, 1125-1142 (1990).
5. C.A. James, R.J. Simmonds and S.A. Wood. Development of an efficient HPLC analysis for Lincomycin for use in contract laboratories, in: "Methodological Surveys in Biochemistry and Analysis", Volume 20, E.Reid and I.D.Wilson (eds.), Royal Society of Chemistry, Cambridge (1990).
6. S.A. Wood, S.A. Rees and R.J. Simmonds. Analysis of brunaprolast, an esterase unstable drug, with an active metabolite subject to oxidative degradation, in blood plasma. *J. Liquid Chromatog.* 13, 3809-3824 (1990).
7. C.A. James, R.J. Simmonds and N.K. Burton. An HPLC assay for a prostacyclin analogue, ciprostene calcium, in human plasma. *J. Liquid Chromatog.* 13, 1143-1158 (1990).

THE ANALYSIS OF PRIMARY AND SECONDARY FREE AMINO ACIDS IN BIO-
LOGICAL FLUIDS: A COMPLETELY AUTOMATED PROCESS USING ON-LINE
MEMBRANE SAMPLE PREPARATION, PRE-COLUMN DERIVATISATION WITH
O-PHTHALALDEHYDE, 9-FLUORENYL METHYL CHLOROFORMATE AND HIGH
PERFORMANCE LIQUID CHROMATOGRAPHIC SEPARATIONS

J.D.H. Cooper[1], D.C. Turnell[1], B. Green[1], D. Demarais[2] and P. Rasquin[2]

[1] Clinical Innovations Limited
Barclays Venture Centre
University of Warwick Science Park
Sir William Lyons Road
Coventry. CV4 7EZ. UK

[2] Gilson Medical Electronics
72 rue Gambetta
Villiers-le-Bel
France

SUMMARY

A fully automated technique for the analysis of amino and imino acids allowing
direct sampling of raw biological fluids is described. The procedure uses on-line membrane
(ASTED^{TM}) technology to remove macromolecular interferences from samples, simultaneous pre-column derivatisation with o-phthalaldehyde and 9-fluorenylmethyl chloroformate
and separation of the analytes using reversed-phase high performance liquid chromatography. Associated problems of both pre-column derivatisation reagents are also overcome. The
use of membrane clean-up also permits an examination of the technique to automatically
estimate non-protein bound tryptophan. Method validation includes method sensitivity, the
use of enrichment to extend the analytical range and correlation with ion exchange chromatographic methodology.

INTRODUCTION

The need to separate amino acids initiated the chromatographic pioneering work of
Martin and Synge [1]. With the plethora of amino acids and metabolites present in a variety
of biological, industrial and environmental samples it is not surprising that the analysis of
such compounds has received a great deal of attention over the last 50 years. The comprehensive review by Deyl et al [2] states that in the biological and clinical sciences there are
nearly two hundred compounds that have been classified as amino acids as opposed to the 20
essential and non essential amino acids recognised for protein biosynthesis.

Sample Preparation for Biomedical and Environmental Analysis,
Edited by D. Stevenson and I.D. Wilson, Plenum Press, New York, 1994

Although it is common practice to estimate free amino acids and metabolites in samples such as plasma and urine, for the examination of inherited metabolic disorders and parenteral nutrition, clinical laboratories have analysed many other types of sample for amino acids [3-7]. This requirement, and the diversity of compounds under investigation, amplifies the necessity for highly specific methods. The early ion exchange chromatography techniques using post column derivatisation with ninydrin [8,9] have led to the identification of a multitude of ninhydrin positive compounds in biological fluids and it could be argued that ninhydrin is relatively non-specific reacting with numerous species other than amino acids [10]. Although the ion exchange techniques are still a valuable analytical tool, the complexity of amino acid analysis due to the diversity of sample types, and varying combinations of amino acids, has exposed the limitations of post column reaction procedures. Much attention, then, has been focussed on reversed-phase high performance liquid chromatographic (HPLC) techniques using pre-column derivatisation methods to improve sensitivity, specificity and speeds of separation. Numerous types of derivatising agents have been used for this mode of operation [11-14] but it would appear that no one reagent is ideal for all the combinations of amino acids and sample types. The pre-column derivatising agent that has attracted most attention is o-phthalaldehyde (OPA), since its reaction with the primary amine of the amino acid in the presence of a thiol (typically 2-mercaptoethanol, MCE) is rapid under aqueous alkaline conditions. Furthermore the fluorescent yields of the alkyl thio substituted isoindole products [15] of this reaction are extremely high. Since its introduction by Roth [16] in 1971, the relatively low polarities of the OPA/MCE amino derivatives have resulted in many publications for the rapid chromatographic separation of these molecules on reversed-phase HPLC [17-22].

Unfortunately, although the OPA/MCE reaction possesses many valuable attributes, it suffers from several limitations, some of which have now been overcome: a few of the OPA/MCE amino acid derivatives have been shown to be unstable due to molecular rearrangement of the isoindole products [16,23]. Whilst substituting MCE with other thiols e.g. mercaptopropionic acid [24,25] has demonstrated improved stabilities of the OPA amino acid derivatives, it has also been shown that those more unstable OPA/MCE amino acid fluorophores do not produce analytical imprecision when the reaction is automated [26-29]. Moreover, the derivatives once on column do not degrade further [30]. The poor fluorescence of the OPA/MCE sulphydryl amino acids has been overcome by the use of the alkylating agent, iodoacetate (IDA) [31,32]. This reacts with the sulphydryls on reduced cystine to produce the compound, S-carboxymethylcysteine, which reacts with OPA/MCE to give an isoindole with approximately the same fluorescent intensity as non-sulphydryl amino acids. It has also been shown that when ammonia is present at high concentrations, the OPA/MCE ammonia derivative can interfere with the chromatographic separation of the OPA/MCE amino acid fluorophores. Complexation of ammonia with tetraphenylboron (TPB) has surmounted this problem [33].

By far the most critical problem of OPA is its lack of reaction with secondary amines. Two approaches have been used to overcome this. Firstly by chemically modifying the imino acid to yield a primary amine which can then be derivatised with OPA/MCE and secondly the use of a further derivatising reagent in combination with OPA/MCE. The former approach has proved difficult. A pre-column derivatisation method has been developed combining chloramine T and borohydride treatment prior to the OPA/MCE reaction that gives good fluorescent yields of both amino and imino acids [34,35]. However, competition between the oxidising agent and OPA/MCE for amino acids occur and the reaction is extremely difficult to automate. Of the reagents available that can be combined with the

OPA/MCE reaction or used alone, 9-fluorenylmethyl chloroformate (FMOC) appears to be the most suitable, reacting rapidly with primary and secondary amines in alkaline conditions. Originally FMOC was developed as a protective group in peptide synthesis [36] but has since been used as a single derivatising agent for amino and imino acid analysis [37,38] forming products with similar fluorescent yields to those of OPA/MCE amino acids. FMOC derivatives of amino acids are also suitable for reversed-phase HPLC separations. As an amino acid derivatising agent, FMOC also has limitations. Like OPA/MCE, poor fluorescence yields are obtained with sulphydryl amino acids. Since FMOC is an acid chloride it is highly reactive to numerous functional groups other than amines e.g. thiols, alcohols [39] and water. Also multiple derivatives of some amino acids like tyrosine [37] and histidine [40] can be obtained. FMOC and its alcohol hydrolysis product both have strong fluorescence at the same excitation and emission wavelengths to the amino acid derivatives. Both compounds can thus interfere chromatographically. Furthermore since FMOC is insoluble in water it is necessary to dissolve the compound in organic solvents. To obviate these problems, organic extraction has been used to remove the excess FMOC and its hydrolysis product. In addition to the difficulties of automating this extraction procedure, losses of some amino acid derivatives also occur [40]. To reduce these complications it is necessary to reduce the FMOC concentration but this imposes limitations on the reaction kinetics with a consequent loss of flexibility of sample volumes and sensitivity when derivatising untreated samples. A further approach has used the reaction of excess FMOC with 1-aminoadamantane [40] to move the interfering peak of FMOC to a more remote region of the chromatogram where amino acid derivatives do not elute. However, the problem of the FMOC hydrolysis product still remains. A comparison of OPA and FMOC and reactions, demonstrating the complementary factors that exist between the two reagents is shown in Table 1. Although the combination of OPA and FMOC amino acid derivatisation has been reported [39,41] for the simultaneous chromatographic separation of primary and secondary amino acids and rapid separations of imino acids alone (by exploiting the different excitation and emission wavelengths for OPA and FMOC derivatives) no attempts to obviate all of the associated problems combined with the automatic preparation of untreated biological samples have been described. The preparation of biological samples poses further problems. Due to the complex nature of biological sample matrices, it has always been necessary to deproteinise or remove particulate contaminants to protect the analytical column whether using pre- or post-column derivatisations. The usual methods for sample preparation include chemical protein precipitation [17,19] and ultrafiltration [42]. Both approaches have proved extremely difficult to automate on line, without using sophisticated and expensive robots.

Pre-column derivatisation of amino acids with OPA/MCE combined with sample preparation using on-line membrane de-proteinisation has been automated [43]. From this beginning the ASTED^{TM} (automated sequential trace enrichment of dialysates) system using cartesian robotics and incorporating the facility for five different pre-column chemistries in combination with on-line sample de-proteinisation was developed providing a useful analytical tool for the complete automation of a variety of analyses [44-50]. This paper represents the culmination of several years development for the analysis of amino acids and demonstrates the flexibility of the ASTED system to completely automate the pre-column derivatisation of amino and imino acids with the combined OPA/FMOC reaction. At the same time the system not only overcomes the associated problems of the reactions, but the filtration potential of the membrane also removes excess FMOC, the insoluble ammonia-TPB complex and high molecular weight contaminants. Descriptions for different HPLC gradients and examples of chromatograms to encompass the complete automatic amino acid analysis for a variety of amino acid compositions are also presented.

Table 1.   Comparison of the OPA and FMOC reagents for amino/
imino acid derivatisation.

Reaction Specification	OPA	FMOC
Reaction conditions	Aqueous Alkaline	Organic Alkaline
Specificity for amino acids (-NH2) only	Good	Poor
Reaction with amino acids (-NH)	No	Yes
Reaction with amino acids	Yes	Yes
Formation of more than one product	No	Yes
Stability of derivatives	Some poor	Good
Reaction with sulphydryl amino acids	Poor	Poor
Ease of automating derivatisation	Good	Poor
Interference by excess ammonia	Yes	Yes

EXPERIMENTAL

Instrumentation

Unless otherwise stated, all the equipment was obtained from Gilson Medical Electronics, Villiers-le-Bel, France.

HPLC Unit

The HPLC equipment consisted of a binary or ternary gradient system incorporating 305/5SC pumps. Either one or two, 121 fluorimeters, connected in series, were used. For the OPA/MCE derivatives the excitation and emission wavelengths were 230 nm (excitation) with a cut-off of 417 nm (emission) respectively. For the FMOC derivatisation the excitation and emission wavelengths were 265 nm and a broad band filter of 300 to 390 nm. Control of the gradient HPLC system and integration of chromatographic peaks was made using a 714 system controller (IBM AT with hard disc, EGA graphic card, mouse, MS DOS 3.1 and Windows software). The HPLC column was 150 mm x 4.6 mm i.d. containing Rainin microsorb 5 µm ODS (Anachem Ltd., Luton, U.K.). No guard columns were used.

Sample Preparation (ASTED) Unit

The ASTED unit has been described in a previous publication [51]. Unless otherwise stated the 7010 Rheodyne injection valve was fitted with a 100 µl loop. The dialyser unit was a Kel-F type with a 100 µl donor and 175 µl recipient chamber volume fitted with the standard cuprophan membrane having a molecular weight cut-off of 15 kDaltons. The sample preparation process was controlled by ASTED version 1.2 software, and unless otherwise stated, configured to PROCESS 2 [51]. This utilises injection of dialysates onto the analytical column via a standard injection loop. The software allows operating parameters e.g. sample and reagent volumes and reaction times to be varied independently. The system was operated in a sequential manner allowing the preparation of the sample and derivatisation immediately prior to injection onto the HPLC column. The configuration of the system is shown in figure 1.

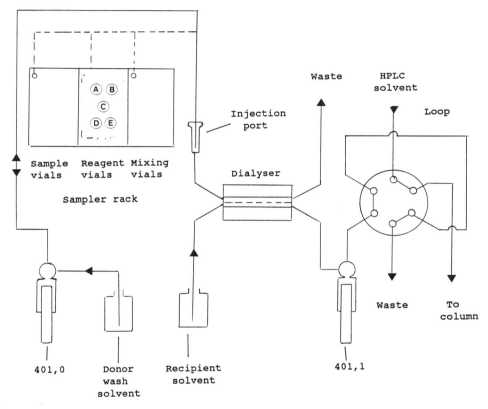

Figure 1.   The configuration of the ASTED system for pre-column derivatisation of amino acids. The arrows indicate direction of flow and the valve is in the LOAD position.

Ion Exchange Unit

For comparison of the HPLC amino acid results, ion exchange chromatography was performed. This was carried out on a BiotroniK LC5001 (Biotech Instruments, Luton, U.K.) amino acid analyser, incorporating a 170 mm glass column packed with BiotroniK BTC 2710 resin, and using an extended physiological program recommended by the manufacturer. After reaction with ninhydrin, amino acid derivatives were detected by absorbance at 440 and 570 nm.

Reagents

All HPLC solvents were obtained from BDH Chemicals Ltd., Poole, U.K. Unless otherwise stated all other chemicals were obtained from Sigma, Poole, U.K. The water used for all reagent preparations and ASTED operations was purified through activated carbon and an ion exchange resin (Spectrum C system, Elga, High Wycombe, U.K.), and passed through a 0.2 μm filter. A 0.2 M borate buffer, pH 9.5, was prepared.

Amino Acid And Metabolite Standard Solutions

The range of amino acids, metabolites, imino acids and concentrations employed for the investigation is shown in Table 2. The amino acids were dissolved in 0.1 m hydrochloric acid. Amino acid solutions were prepared individually or as a combined mixture of all those listed. Stock solutions were normally diluted 1 in 10 in borate buffer.

Derivatising Reagents

Internal Standard (IS)/MCE/TPB Reagent: 1.0 ml of a stock solution (in water) containing 20 mmol/l each of homocysteic acid (HCA; IS for amino acids), homoserine (HSER), norvaline (NVAL) and 50 mmol/l thioproline (TPRO; IS for imino acids) was diluted to 25 ml with borate buffer and 100 μl of MCE added. Varying amounts of sodium tetraphenylboron (BDH Chemicals Ltd.) were added to 20 ml aliquots of the IS/MCE reagent. The range examined was based on a previous report [33].

Iodoacetate (IDA) Reagent: 3.5 g of IDA (99% purity) was dissolved in 50 ml of borate buffer. The pH was adjusted to 9.5 with 4 m sodium hydroxide solution and the volume made to 100 ml with the borate buffer.

O-phthalaldehyde/2-mercaptoethanol (OPA/MCE) Reagent: The composition of the OPA/MCE reagent was based on a previous report [29]. Varying amounts of OPA (Sepramar grade BDH, Chemicals Ltd.) were dissolved in 10 ml of methanol and made to 100 ml with borate buffer, followed by the addition of MCE (4 parts of MCE:1 part of OPA).

Fluoroenylmethyl Chloroformate (FMOC) Reagent: Varying amounts of FMOC were dissolved in 20 ml aliquots of dry acetone. These reagents were observed to be stable for one week at ambient temperatures.

Samples

Unless otherwise stated samples were either plasma (heparinised), serum or urine. All sample types were stored frozen at -20°C. After thawing the mixed plasma was placed in a sampler vial and analysed automatically without further manual treatment. For ion exchange methods plasma samples are usually precipitated within 30 mins after collection due to losses with time of the sulphydryl amino acid, cystine. No loss of cystine occured when using the ASTED/HLPC method [52]. Urine samples, acid hydrolysed (with 6M hydrochloric acid) and dessicated to remove the acid were also prepared.

Table 2.   Amino/Imino acid stock calibration standard.

No.	A/Imino Acid	Abbreviation	mmol/l
1	Phosphoserine	SER	5
2	Aspartic acid	ASP	5
3	Glutamic acid	GLU	5
4	Arginosuccinic acid	ASA	5
5	Cystine	CYS	5
6	α-Aminoadipic acid	AAD	5
7	Asparagine	ASN	5
8	Homocystine	HCYS	10
9	Serine	SER	5
10	Histidine	HIS	5
11	Glutamine	GLN	5
12	Phosphoethanolamine	PEA	10
13	Glycine	GLY	5
14	Threonine	THR	5
15	Citrulline	CIT	5
16	Arginine	ARG	5
17	Carnosine	CARN	10
18	3-methylhistidine	3-MH	5
19	β-Alanine	ALA	5
20	Homocitrulline	HCIT	5
21	Alanine	ALA	5
22	Taurine	TAU	5
23	β-aminoisobutyric acid	BAIB	5
24	τ-aminobutyric acid	GABA	5
25	Tyrosine	TYR	5
26	α-aminobutyric acid	ABA	5
27	Ethanolamine	ETH	5
28	Valine	VAL	5
29	Methionine	MET	5
30	Tryptophan	TRY	5
31	Cystathionine	CYST	10
32	Phenylalanine	PHE	5
33	Isoleucine	ILEU	5
34	Leucine	LEU	5
35	Hydroxylysine	HLYS	10
36	Ornithine	ORN	10
37	Lysine	LYS	10
38	Hydroxyproline	HPRO	10
39	Sarcosine	SARC	10
40	Proline	PRO	10
41	Homocysteic acid	HCA (IS)	
42	Homoserine	HSER	
43	Norvaline	NVAL	

Table 3. Solvent compositions for three gradient mobile phases described. All numbers are percentages of total volume.

HPLC Solvents	Ternary Gradient for Physiological Samples	Binary Gradient for e.g. Hydrolysates	Binary Gradient for Secondary amines
Solvent A			
Sodium Phosphate (250 mmol/l)	15.0	15.0	23.0 (pH 5.6)
Propionic Acid (250 mmol/l) pH to 6.6	20.0	20.0	
Acetonitrile	7.0	7.0	
Dimethylsulphoxide	3.0	3.0	
Methanol			11.0
Water	55.0	55.0	66.0
Solvent B			
Acetonitrile	40.0	40.0	40.0
Methanol	33.0	33.0	40.0
Dimethylsulphoxide	7.0	7.0	
Water	20.0	20.0	20.0
Solvent C			
Sodium Phosphate (250 mmol/l)	25.0		
Propionic Acid (250 mmol/l) pH to 5.5	20.0		
Acetonitrile	7.0		
Dimethylsulphoxide	3.0		
Water	45.0		

Chromatography Conditions: All HPLC solvents were filtered through a 0.45 µm filter (Millipore Ltd., London, U.K.) and degassed with helium. One of three different sets of gradient conditions (table 3) was used depending on the number of amino acids under investigation. In each of the sets, solvent 'A' contained various mixtures of 0.25 m solutions of anhydrous disodium hydrogen orthophosphate and propionic acid at different pH's. The gradient programs employed are shown in table 4. The time for column re-equilibration was 5 min and the solvent flow rate used was 2.5 ml/min. The column was used at ambient temperature.

Another binary gradient solvent was employed to examine the use of trace enrichment in the system. This incorporated 100% solvent A rising to 100% solvent C (of the ternary gradient composition described, table 3) in 10 min at a solvent flow rate of 2.5 ml/min.

Table 4. Programs for the three gradient applications described.

Time (mins)	Ternary Gradient for Physiological Samples %B	%C	Binary Gradient for e.g. Hydrolysates %B	Binary Gradient for Secondary amines %B
0.0	0.0	0.0	0.0	45.0
3.5			0.0	
5.0			9.0	
7.9				45.0
8.0	0.0	0.0		
8.1				100.0
9.5			9.0	
10.8				100.0
10.9				45.0
12.0			28.0	
17.0			35.0	
19.0	8.0	0.0		
19.1	13.3	86.7		
20.0			55.0	
20.2			100.0	
22.5			100.0	
22.7			0.0	
27.0	27.0	73.0		
27.5	28.0	0.0		
29.0	32.0	0.0		
34.5	45.0	0.0		
39.0	100.0	0.0		
39.6	100.0	0.0		
40.0	0.0	0.0		

ASTED Conditions: Unless otherwise stated the solvents in the 401 dilutor 0 and 1 reservoirs (Figure 1) contained deionised, filtered water.

PROCEDURES

HPLC/ASTED Sample Preparation And Derivatisation

Total Amino And Imino Acid Estimation. The derivatisation sequence is shown in Table 5 and the location of sample vial, derivatising reagents and mixing vials is shown in Figure 1. After mixing the sample with each reagent in turn the mixture was loaded into the dialyser. Both the donor stream (sample/reagent mixture) and the recipient stream were then held static for a period of 2.5 min to allow diffusion of the derivatised amino and imino acids. After this time the recipient stream (dialysate) was moved using the 401 dilutor 1 into the injection loop of the Rheodyne valve and the amino acid derivatives injected onto the HPLC column. With the Rheodyne valve in the inject position both the donor and recipient channels were purged and the system was then ready to process the next sample.

Table 5. The derivatisastion reaction sequence used for total amino/imino acid analysis.

Sample Preparation	Reaction		Time
Take 20μl sample and mix with:			
50 μl MCE/IS/TPB solution	Cystine NH3 + TPB	Cysteine NH3-TPB complex	
50 μl IDA solution	Cysteine	S-carboxy-methylcysteine	
50 μl OPA/MCE solution	R-NH2 + OPA	Fluorophore	
20 μl FMOC reagent	R-NH + FMOC	Fluorophore	5 mins
Load Reaction Mixture into donor channel of dialyser			
Wait	Allows diffusion of fluorophores into recipient channel of dialyser		
Move recipient stream into injection loop			
Injection	Chromatography and detection of fluorophores		
Purge system with 4000 μl			40 mins
Make report and regenerate gradient			
Take next sample			

Free Amino Acid Estimation

The system had the same configuration to that used for total analyte estimation (Figure 1). However an OPA/MCE solution (0.004 m OPA, 0.015 m MCE in 0.2 m borate buffer solution, pH 9.5) was placed in the recipient reservoir of the dialyser. The donor channel of the dialyser was filled with 100 μl of sample using the 401 dilutor 0. The dialysis times, injection and system purging were completed in an identical manner to the total amino acid estimation by elution using the mobile phase.

Trace Enrichment Of The Dialysates

The ASTED unit was configured for PROCESS 1. This incorporated the use of a trace enrichment cartridge packed with 70 mg of 10 μm Hypersil ODS (Shandon Southern Products, Runcorn U.K.) and substituted for the loop on the Rheodyne injection valve. This process has previously been described in more detail [51]. The derivatisation was carried out in the manner described for total amino acid analysis but omitting the FMOC derivatisation by simple alteration of the ASTED software reagent variables. However, after loading the reaction mixture into the dialyser, the derivatised sample was held static whilst the recipient solution was moved continuously (at 1.0 ml/min) into the trace enrichment cartridge. After 4 min the Rheodyne valve was switched to the INJECT position and the enriched OPA/MCE amino acid derivatives injected onto the column by elution using the mobile phase.

Quantification

Quantification and peak identification was made using the Gilson 714 software. Amino acid derivatives were identified by their retention times relative to the reference peaks produced by homocysteic acid, homoserine, norvaline and imino acids and sarcosine by their retention time relative to the reference peak, thioproline. The amino and imino acid concentrations were quantified by a comparison of their peak areas with that of the appropriate IS (homocysteic acid or thioproline) using the internal standard mode of the Gilson 714 integration software. On occasions, for experimental investigations only, integrated peak areas were used.

Sample Preparation For The Ion Exchange Method

Serum samples were de-proteinised with sulphosalicylic acid (SSA) using the method recommended by the manufacturer of the amino acid analyser. Peak areas were calculated by multiplying the peak height and width at half peak height.

Experimental Investigations

Unless otherwise stated the total amino/imino acid derivatisation procedure and ternary gradient conditions were used for these investigations and the reactant concentrations are those in the final reaction mixture.

Optimisation Of Reactant Concentrations

The optimisation of the reaction conditions was examined by individually varying the concentration of each reactant. A pooled serum and urine, each supplemented with 1.0 mmol/l of different amino acids representative of a range of OPA/MCE amino acid polarities and FMOC imino acid derivatives were analysed for the optimisation. The amino acids used included glutamate, serine, alanine, taurine, tryptophan, lysine and proline. An aqueous standard containing 1.0 mmol/l of the same amino/imino acids was analysed together with the supplemented serum and urine samples. This was performed for a range of OPA (1.0 to 24 mmol/l), FMOC (1.0 to 10.0 mmol/l) and TPB (0 to 30.0 mmol/l) concentrations. Further to this a range of aqueous ammonium chloride concentrations (0 to 500 mmol/l), each containing 1.0 mmol/l of proline, were analysed with and without the presence of optimum TPB concentrations and using the binary gradient conditions for imino acid analysis (tables 3 and 4). During the investigation of each reactant, the concentrations of the others remained constant and are identified on the appropriate figure legends. The optimum concentration of MCE for the OPA reaction [30] and IDA [31] has been established in previous communications and when derivatising untreated samples for this investigation minimum experimentation was carried out to ensure no detrimental effects occurred on the alkylation of the reduced cystine when including FMOC into the reaction scheme. This was performed by assaying the FMOC assay for imino acids with and without the presence of IDA using the samples prepared for the optimisation of the reaction.

Matrix effects

The magnitude of matrix interference was determined using three different procedures as follows:-

1. Firstly by the recovery of amino and imino acids from specimens having various concentrations of serum (protein). This was performed using the optimum reaction conditions determined. The pooled serum used was first assayed for the amino/imino acid concentrations and then diluted with water to give a range of serum (protein) concentrations. Each dilution was then supplemented with 1.0 mmol/l of amino/imino acids identified in the optimisation investigations. An aqueous standard containing 1.0 mmol/l of the same amino acids providing a protein dilution range of 0 to 100% (assuming no dilution effect during supplementation of the serum). After chromatographic separation the peak areas (multiplied by the appropriate serum dilution factor) of the identified amino/imino acids in the non-supplemented serum were subtracted from the relevant peak areas of the supplemented serum at the various serum dilutions. These results were expressed as a percentage of the aqueous standard.

2. By analysing 20, 40 and 60 µl urine and serum sample volumes using optimised reagent conditions. No changes to reagent volumes were made. The peak areas of each amino/imino acid derivatives were compared as well as the ratio of the peak areas to the internal standard, homocysteic acid.

3. For the amino acid tryptophan, the free amino acid derivatisation procedure was adopted. The sample dilution treatment was performed exactly as described in the matrix recovery experiment. No attempts were made to adjust the pH of the pooled serum being used, prior to analysis, but after dialysing against the OPA/MCE reagent the emerging (spent) serum was captured and the pH measured.

Sensitivity Of The Method Procedure

A number of variables can create changes in the assay sensitivity. The efficiency of the ASTED system (i.e. the absolute recovery of the amino acid derivatisation) was examined using only the OPA/MCE procedure since, it was necessary to directly inject derivatives onto the column (impractical with high concentrations of excess precipitated FMOC). To perform this estimate the peak area resulting from known amounts of amino acids was determined by derivatising an aqueous standard solution and injecting directly onto the HPLC column using a filled loop (20 µl) technique with the 7010 Rheodyne valve. The same standard solution was then analysed using the ASTED derivatising system described for OPA/MCE amino acids alone. The absolute recovery was calculated by comparing the peak areas of the two analyses and taking into account the donor volume of the dialyser. To examine the factors affecting sensitivity two method variables received particular attention:

-   Dialysis diffusion rates were examined by analysing an aqueous standard and pooled serum sample. After derivatisation and loading the donor channel of the dialyser, the reagent and sample mixture was retained for a range of dialysis times (20 to 360 s). The dialysis times were varied by altering the appropriate time variable in the ASTED software.

-   An aqueous standard was analysed for amino acids only (omitting the FMOC derivatisation) with the fluorescence detector range set at 0.1. This standard was diluted 1 in 100 with 200 mmol/l borate buffer (pH 9.5) and re-analysed with the fluorescence detector range set at 0.001. This solution was further diluted 1 in 50 with the borate buffer and re-analysed using the ASTED derivatisation system configured for the trace enrichment procedure described and the fluorimeter set at a detection range of 0.001.

Validation Of The Assay

Chromatography And Practical Examples

The chromatographic resolution was examined for various sample types: for the ternary gradient an aqueous standard, serum, urine, haemolysate, homogenised brain tissue (white matter); for the binary gradient of essential and non essential amino/imino acids an aqueous standard; for the binary gradient of imino acids only, an aqueous standard and a hydrolysed urine sample.

Comparison Of The ASTED/HPLC And Ion Exchange Methods

The assay procedure was validated by comparing the amino acid concentration results of 20 fresh plasma samples obtained from healthy volunteers and analysed by the ASTED/HPLC and the ion exchange method.

Analytical Performance

Using the ternary gradient system and the total amino/imino acid derivatisation method described, the within batch imprecision of the technique was estimated by assaying 30 pooled serum and 30 pooled urine samples supplemented with the stock calibration standard to achieve two amino/imino levels (within the analytical linear range). An aqueous standard was analysed every 5 samples. The between batch imprecision was estimated from 15 replicate analyses of two serum and two urine samples analysed in 30 separate runs over a 30 day period. In this case samples were stored at -20°C. During the within batch study the original unsupplemented samples were also analysed to establish the analytical recovery of the analytes from the biological matrices.

Chromatographic And Sample Preparation Interferences

The ability of the ASTED system to remove excess FMOC and its hydrolysis product was ascertained by comparing the peaks obtained for a direct injection of 50 μmol/l of FMOC in water and for the preparation of a 5 mmol/l FMOC solution in water using the ASTED derivatisation procedure described. During the development of this technique other compounds (e.g. antibiotics) were analysed that have been observed to interfere chromatographically with the classical ion exchange technique.

RESULTS AND DISCUSSION

The analysis of amino and imino acids in biological fluids is a difficult problem. Strategically complete automation of amino acid systems is required considering the complexity of the derivatisation, sample preparation, and chromatographic separations that are necessary to achieve the analytical goals. For the analysis of amino and imino acids in biological samples, whatever their type, it is essential to derivatise the analytes to obtain the necessary specificity. For rapid amino acid analysis using reversed-phase separations, pre-column derivatisation with OPA has gained most popularity due to the fact that the reaction is relatively easy to automate. It could be argued that OPA is the most specific reagent available for amino acid analysis since unlike other reagents its reaction with even small peptides

[53] is poor under the reaction conditions usually employed. However, since it is only specific for primary amines its lack of reaction with imines has deterred its general acceptance for use in amino acid analysis. FMOC has the potential to overcome this drawback but its nature is entirely different to OPA. It is highly reactive to many other functional groups and the reaction is more difficult to automate. It can be seen in table 1 that the combination of OPA and FMOC tends to overcome the limitations of each reagent.

The development of the ASTED system allowed on-line sample preparation combined with derivatisation and the separation of amino/imino acids derivatives in an automatic mode. In this sense it could be termed an amino acid analyser for this particular application. For this automation to occur several technical problems needed to be addressed: All of the limitations of the OPA/FMOC reaction needed to be overcome (table 1); direct sampling of untreated specimens to remove high molecular weight contaminants was required; followed by pre-column derivatisation of the amino acids with OPA/MCE then the reaction of the imino acids with FMOC and separation of the analytes on reversed-phase columns. Most of these technical problems and investigations are inter-related and required examination in terms of reaction kinetics, the ASTED system and sample preparation, validation of the method and chromatographic detection and resolution.

Optimisation Of Reaction Conditions

The major benefit of pre-column derivatisation compared with post column reactions is that more complex chemistries can be performed providing that chromatographic interferences are not evident. Further specificity can be gained, of course, by the choice of derivatising reagent, detector and the use of different gradient conditions. The flexibility of the hybrid ASTED system [54] provided a number of options for the location of the derivatisation reagents. In order to exploit the full potential of the Gilson 231 robotic sampling device, derivatisation was performed on the donor side of the dialyser (Figure 1), for the majority of applications. Of the other alternatives one is demonstrated using the estimation of free tryptophan in serum as a model. Numerous publications [16,30,37,41] have examined the kinetics and reaction conditions for both the OPA and FMOC mechanisms. However, this has only been carried out after pre-treatment of the samples, using physical or chemical means, to remove, for example macromolecules. When preparing specimens on-line with the ASTED system the untreated samples were derivatised without such pre-treatment. In this case there was the likelihood of other endogenous compounds reacting with the derivatising agents OPA/MCE, and in particular, FMOC. The actual reaction sequence for this derivatisation process is shown in table 5. Both OPA/MCE reactions are optimal at a pH of 9.5. The use of MCE and IDA to reduce and alkylate cystine in untreated serum for acceptable fluorophore production with OPA/MCE has been reported [43]. In this study no effects on the S-carboxymethyl cysteine isoindole produced from this reaction were observed chromatographically when incorporating FMOC into the reaction scheme. Furthermore it is probable that the incorporation of IDA has beneficial effects on the FMOC reaction kinetics, removing excess endogenous and exogenous thiols that will compete with imino acids for FMOC. The fluorescence response (peak area) obtained for the amino acid serine in aqueous standard, supplemented serum and urine samples with increasing OPA concentrations employing the reagent volumes described (Table 5) is shown in Figure 2a. The other amino acids used in this investigation showed similar patterns. In both biological samples some endogenous competition with the amino acid for the OPA can be observed. A final stock OPA concentration of 37 mmol/l was adopted to ensure zero order kinetics. At this concentration of OPA, 144 mmol/l of MCE was added to the stock OPA solution. Similar patterns were observed with varying FMOC concentrations and its effect on the recovery of proline from plasma and urine samples (Figure 2b). To obtain zero order kinetics a stock FMOC concentration of

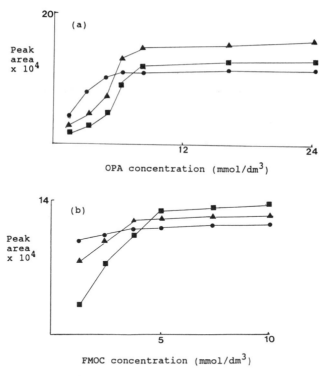

Figure 2. The variation of detector response with (a) increasing OPA concentrations on the amino acid SER; (b) increasing FMOC concentrations on the imino acid PRO in different matrices. The symbols represent ● aqueous standard; ■ serum supplemented with serine; ▲ urine supplemented with serine. The concentration of reactants are those in the final reagent mixture.

50 mmol/l was used. All of the concentrations quoted are derived from the volumes described in the derivatisation procedure (Table 5). For both OPA and FMOC the increased recovery of the amino and imino acid from the supplemented serum and urine sample compared with the aqueous standard (Figures 2a and 2b) was due to endogenous amino/imino acid present in the biological sample. In this procedure acetone was used to dissolve the FMOC and for this reason a Kel-F dialysis unit was utilised in the system. The volume of the acetone/FMOC reagent was minimal to prevent protein denaturation (if present in the sample).

The optimisation of the TPB concentration, when combined with OPA/MCE derivatisation of a urine sample, has been examined previously [33]. However, this was carried out using a manual procedure where the precipitated ammonia-TPB complex was removed by centrifugation. It was considered necessary to re-examine this situation since, in this automated approach, the fine precipitate of ammonia-TPB produced was loaded into the dialyser together with derivatised amino acids. Furthermore the effects of TPB on the FMOC derivatisation of imino acids have never been examined. The effect of TPB on the OPA/MCE amino acid derivatives with the FMOC reagent combined was found to concur with previous experimentation [33]. No loss of fluorescence of any OPA/MCE amino acids was observed but chromatographic interference by the ammonia OPA/MCE derivative was evident without

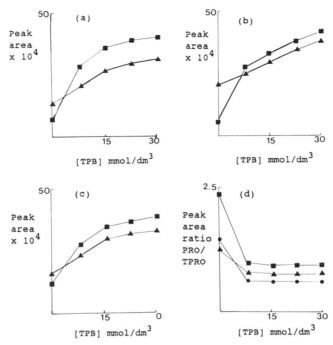

Figure 3. The fluorescence response of ▲ PRO and ■ TPRO with varying concentrations of TPB from (a) aqueous solution; (b) supplemented urine; (c) supplemented serum. Plot (d) shows the peak area ratio of PRO/TPRO with varying TPB concentrations; symbols ● ■ ▲ represent aqueous, urine and serum results respectively. The concentration of TPB is that in the reaction mixture and optimum concentrations of other reactants was maintained throughout.

the inclusion of TPB in the reaction sequence. Furthermore it was apparent that the incorporation of TPB in the reaction had beneficial effects on recovery of the dibasic amino acid (e.g. lysine) OPA/MCE derivatives. These derivatives can be problematical due to their increased lability [30] compared with other amino acid OPA/MCE molecules.

The effect of high ammonia concentrations on the FMOC reaction was found to be far more competitive. Figures 3a,b and c show the fluorescence response of the IS (thioproline) and proline in aqueous solution, serum and urine samples supplemented with proline with increasing TPB concentrations. Increasing fluorescence was observed for both analytes with increasing concentrations of TPB in all three sample types. However the IS was more affected than proline, but at sufficient TPB concentration the ratio of proline peak area to IS peak area became constant (Figure 3d) and ensured precise quantitative analysis. A comparison of the TPB effects on the fluorescence recovery of the imino acids with urine and serum samples showed some differences. That ammonia competes for the FMOC is shown in Figure 4. The fluorescence response for an aqueous proline solution incorporating different ammonium chloride concentrations was compared with and without the addition of TPB using the derivatisation procedure for total amino acid analysis described. These results were obtained using the derived optimum TPB stock concentration of 60 mmol/l and the FMOC IS derivative showed a similar response to that of proline. At the optimum TPB concentration used competition effects of up to 250 mmol/l of ammonium ion can be suppressed with the inclusion of TPB, which is sufficient for urine samples. Although competitive effects of ammonia for FMOC have been demonstrated, the unexpected results for the effect of TPB on aqueous and serum samples (Figures 3a and 3b) suggests that TPB has more beneficial properties than just removing ammonia. As a complexing agent TPB is not specific for ammonia but reacts with other cations [55] that may also interfere with the FMOC reaction.

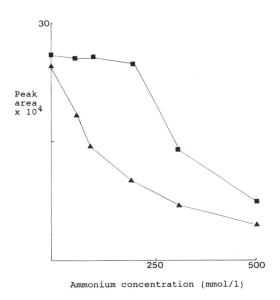

Figure 4.   The effect of increasing ammonium concentration on the recovery of PRO with ■ , and without
▲ , the presence of TPB.

The ASTED System And Sample Preparation

General Considerations. The need to prepare complex biological material prior to chromatography is well recognised [56-58]. Historically for amino acid analysis, whether using pre- or post-column derivatisation, complex samples usually undergo sulphosalicylic acid (SSA) de-proteinisation [9]. The use of such denaturents have a number of disadvantages: the procedures are difficult to automate; changes to any protein secondary and tertiary structure occur; the efficiciency of the process can be poor if the incorrect sample/acid ratio is used [59] and consequently reduced chromatography column life times prevail; for pre-column technology and reversed-phase HPLC, guard columns are essential. However, the time taken to recover high concentrations of amino acid using this type of sample preparation is low when dealing with individual samples.

Compared with conventional preparation techniques ASTED utilises high performance membrane purification (HPMP) for sample treatment [49,51]. Deproteinisation of samples using membrane dialysis techniques have been used for a number of years in analytical continual flow systems [60] and are simple to automate since the membrane can be regenerated between samples. However the efficiency of such systems is poor. The rate of analyte diffusion across membranes is low and consequently the time required to obtain high yields of compounds under investigation can be so great that it prohibits the adoption of such techniques prior to chromatograhic analysis. However, the development of the ASTED membrane system has improved the analyte rate of recovery by increasing the membrane surface area/sample ratio [51] and incorporating trace enrichment to concentrate sub-analysable analyte levels in the dialysates.

The use of membranes with low molecular weight cut-off characteristics has many benefits: the technique is mild and no changes to protein structure occur; no extremes of pH need be used to remove macromolecular contaminants and losses of amino acids such as tryptophan and glutamine (by hydrolysis) were avoided. The efficient filtration properties of membranes removed the encumbrance of guard columns allowing ease of adaptation of the system to smaller bore column applications and potentially, any type of homogenous liquid sample maybe automatically sampled without pretreatment. Furthermore the membranes were regenerated many hundreds of cycles between samples and long analytical column life times were ensured, producing a system cost effective in both labour and materials. Although, at present, membranes do not particularly aid specificity of analyte separation during sample preparation, future trends may alter this position. However, for the ASTED system, the inclusion of trace enrichment and, in the case of amino acids, derivatisation compensates for this situation. Furthermore, for amino acid analysis the high fluorescence intensities of the OPA/MCE and FMOC amino/imino acid derivatives enabled easy detection of analytes from physiological samples, even when maintaining the donor and recipient stream in a static mode and then injecting the dialysate onto the column after a suitable period of time (table 5). In view of the instability of some of the OPA/MCE amino derivatives the ASTED system was operated in a sequential manner. The system can be operated in a concurrent sequential mode [61], where the sample preparation occurs during the previous samples chromatographic separation, provided that times for the preparation process, integration reporting times etc. are known.

Matrix Effects

Analytes in physiological fluids may be protein bound and since it is only the free or non protein bound molecule that diffuses across the membrane, matrix effects may, quantitatively, produce between sample imprecision. Most amino acids are not protein bound and no matrix effects were observed for the amino acids tested with increasing concentrations of protein. These results are represented by the amino acid serine (Figure 5). This was further confirmed when no losses in analyte recovery with increasing serum and volumes used in the assay were observed, although excessive endogenous reactants will eventually cause deviations from zero order OPA and FMOC reaction kinetics. However, tryptophan is known to be protein bound [62]. The protein binding isotherm produced in this investigation (also Figure 5) after dialysing various samples containing different protein concentrations, but equal tryptophan levels, into OPA/MCE reagent on the recipient stream of the dialyser unit, confirmed this observation. During this investigation no changes in the sample pH was observed. The results are in general agreement with previous findings [62] even though no strict adherence to pH control was made, a factor that is recognised to affect the degree of binding of tryptophan [62]. This potential of automatically measuring an index of free analyte concentrations using the ASTED system has been reported [51] and its demonstration using tryptophan as a model further emphasises the flexibility of the ASTED approach to sample preparation. The need to remove matrix effects is paramount when analysing samples for total tryptophan concentrations. Numerous methods are available to decrease the magnitude of protein binding [51] and in the case of tryptophan the sample/reagent dilution factor used was sufficient to negate most of the protein binding and between sample variability for this amino acid was negligible.

Variables Affecting Method Sensitivity

Most analytical methods are confined to a defined analytical range. However, for amino acid analysis the analytical range varies considerably depending upon the sample type. Although the procedures presented relate to clinical investigations the ASTED procedure can accommodate other analytical fields e.g. research, agricultural, industrial etc. The variables that alter the sensitivity of the ASTED amino acid technique are numerous. If individual amino acids are being investigated changes to the gradient program can give improved peak performance. However, changing the variables in the ASTED sample preparation system can effect major alterations in the analytical range. For static dialysis situations and for the total amino acid assay described, the absolute recovery of the method was calculated to be 3.1%. If dispersion of the analytes in the dialysate is considered negligible the

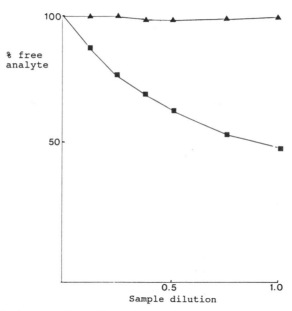

Figure 5.   Protein binding isotherms for ▲ SER, and ■ TRYP. On the X-axis a dilution of 1.0 is equivalent to neat sample.

analyte recovery was much higher since only 100 μl of a total recipient volume of 175 μl was injected onto the column. It is worth noting that smaller sample/reagent volumes can be accurately located in the donor channel if required. Increasing the time during which diffusion occurs will increase the analyte recovery. This is shown in Figure 6 for both OPA/MCE and FMOC derivatives of serine and proline in the aqueous standard. All other amino acids in both the aqueous standard and serum showed a similar pattern. As was expected there was a nonlinear relationship as the analyte concentration in the donor and recipient stream began to approach equilibrium. The rate of analyte transfer across the membrane will alter if the derivatising reagents are placed on the recipient side of the dialyser [43] and is due to back diffusion of the derivatised analytes.

The highest analyte recovery that can be achieved using equilibrium dialysis is 50%. Greater recoveries are achievable using the ASTED system incorporating trace enrichment. This mode of operation enables the recipient stream to be moved continuously and high donor analyte gradient concentrations are maintained [51]. Under the conditions of the assay described it was estimated that approximately 100 pmol of each amino acid OPA/MCE derivative was injected on column (Figure 7a). Even without trace enrichment 1 pmol of amino acid OPA/MCE derivative can be detected on column (Figure 7b). However the inclusion of trace enrichment into the ASTED amino acid system can extend the analytical range considerably. The results for the treatment of a dilute aqueous standard, using combined dialysis and trace enrichment, for a small range of amino acids is shown in the chromatogram in Figure 7c. The concentration of the amino acid OPA/MCE derivatives was equivalent to quantifying 50 fmol on column. Further work is required to establish the chromatographic separations when analysing biological samples using combined dialysis and trace enrichment procedures. One major benefit of incorporating trace enrichment is that it enables the use of larger dialyser units extending the analytical range.

Method Validation

Chromatographic Detection and Resolution. Previous publications [17-22] have emphasised the speed of chromatographic separation using pre-column derivatisation and

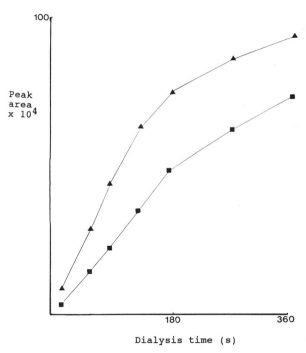

Dialysis time (s)

Figure 6.   The effect of dialysis time on the recovery of ▲ SER OPA/MCE derivative: ■ PRO FMOC
derivative.

reversed-phase HPLC for the analysis of amino acids. The three gradient conditions listed in this report were designed to encompass a variety of applications. In no way can they be construed as definitive since other workers may demand different requirements of an amino acid assay. The final choice of thiol used in the OPA reaction is one of chromatographic expertise since minor changes in the groups of the thiols will cause major changes in the polarities of the OPA/thiol amino acid molecule. Since the problem of instability of the some of the OPA/MCE molecules had been addressed, the incorporation of MCE was continued to incorporation of MCE was continued to avoid unnecessary changes to the mobile phase solvents. However, sample preparation with ASTED would operate equally well regardless of the type of thiol used.

The more complex ternary gradient system was based on an earlier communication [17] which reported the use of binary solvent conditions. This paper described the solvent factors (e.g. buffer concentrations, organic mobile phase modifiers etc.) necessary to effect adequate resolution of many of the OPA/MCE amino acids listed in Table 2. However, the use of ternary solvent conditions was necessary to resolve taurine, β-amino isobutyric acid and τ-aminobutyric acid by virtue of their different pKa's (Figure 8a; the peak identification can be made by reference to Table 2). The column capacity factors (k^1) for those amino acids containing ionisable functional groups (e.g. phosphoethanolamine) were most affected by changes in pH (decreasing pH increased the k^1). The separation of the FMOC derivatives (Figure 8b) was governed entirely by the HPLC solvent conditions employed for the resolution of the OPA/MCE derivatives. However, reducing the pH will increase the k^1 of the imino acid FMOC derivatives. pH changes did not alter the k^1 of excess FMOC and its hydrolysis product which were well resolved under these conditions from the analytes of interest. Sarcosine reacts with FMOC and was included since it has clinical importance [63]. The peak shape of proline was broad compared with other imino acids. This has been recognised [37,38] and is thought to be due to some steric changes of the derivatised molecule but does

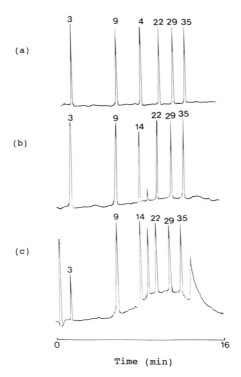

Figure 7.   Chromatograms (a) 100 pmol of each amino acid OPA/MCE derivative on column (detector
attenuation 0.1, dialysis alone); (b) 1 pmol of each amino acid OPA/MCE derivative on column
(detector attenuation 0.001, dialysis alone); (c) 50 fmol of each amino acid OPA/MCE derivative
on column (detector attenuation 0.001, dialysis and trace enrichment). For peak identification
refer to Table 2.

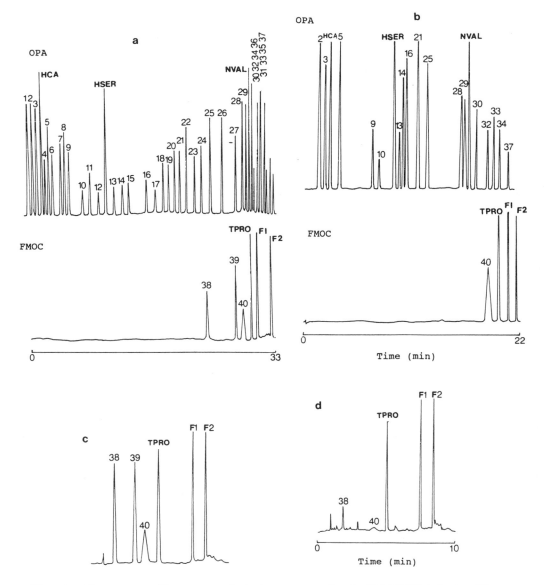

Figure 8.   Chromatograms (a) aqueous standard using ternary gradient for OPA/FMOC derivatives (b) aque
ous standard using binary gradient for OPA/FMOC derivatives (c) aqueous standard using binary
gradient for FMOC derivatives (d) hydrolysed urine sample using binary gradient for FMOC derivatives. Peaks F1 and F2 are FMOC hydroysis product and excess FMOC respectively.

not influence quantification. For the ternary and binary chromatograms shown three reference peaks were included to prevent peak mis-identification due to any gradient deviations. The peak retention times, either between runs or between changes of column varied less than 6% provided that ambient temperatures were kept constant (± 1°C).

The versatility of the ASTED unit for preparing OPA/MCE derivatives of amino acids in a variety of untreated sample types has been demonstrated in a previous publication [51]. The preparation of similar sample types (e.g. haemolysate, homogenised brain tissue) using simultaneous OPA/FMOC derivatisation to that described in this report performed in a similar and acceptable analytical manner. Acid assays are required then simple elimination of the FMOC reagent by the ASTED system can be performed. Contrary to this the different excitation and emission wavelengths of the OPA/MCE and FMOC derivatives can be used to advantage. Elimination of amino acid interferences by reaction with OPA/MCE, on the detection of the FMOC imino acids alone, is shown for the aqueous standard, in Figure 8c. In this way more rapid chromatographic separations can be managed due to the smaller number of analytes that need to be resolved. Clinically, this approach is of benefit for the analysis of total hydroxyproline in urine [64] and the chromatogram for this application is shown in Figure 8d.

As stated the combination of OPA/MCE and FMOC derivatisation can be advantageous if only the analysis of imino acids is required. However, one criticism of the technique

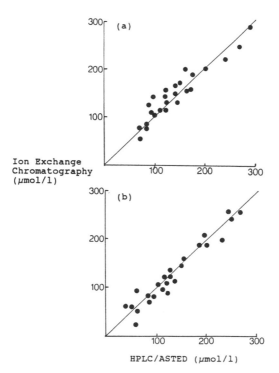

Figure 9. The correlation of ion exchange and HPLC/ASTED methods for plasma amino acids (a) threonine (Y = 1.21X + 8.76, r = 0.946); (b) proline (Y = 1.00X + 9.2, r = 0.953)

described is the combination of two fluorescence detectors connected in series to simultaneously monitor the emerging OPA/MCE and FMOC derivatives from the column. This may be considered as advantageous since the imposition of excess FMOC and its hydrolysis product on the complex chromatographic separations is not such a problem to overcome. Furthermore rapid chromatographic separations can be maintained. As a compromise, preliminary investigations using excitation and emission wavelengths that can accommodate the detection of both OPA/MCE and FMOC derivatives together, without seriously affecting fluorescence intensity have been successful. Further work is required to identify the problems of this approach and to define the mobile solvent phase required to incorporate the combined separation of OPA/MCE and FMOC derivatives.

Correlation Of The ASTED/HPLC With Ion Exchange. The pre-column OPA/MCE method has been shown to give good agreement with the post-column ion exchange technique [19] for plasma amino acids. However, this used a manual de-proteinisation technique and sulphydryl amino acids were not examined. Furthermore, an external quality assessment scheme also demonstrated reasonable agreements between the two techniques [67]. Gross differences between the two procedures was demonstrated for cystine which was later shown to be due to the different sample preparation protocols [52] (protein precipitation methods should be avoided if possible). During this investigation the amino acid homocystine showed identical characteristics to cystine and care should be exercised on its examination in plasma or serum samples.

Figure 9 shows the correlation (using the orthogonal regression of the second kind [68]) of the two techniques for amino/imino acids representative of the total number of analytes (other than sulphydryl amino acids) under investigation. The reasonable agreement between the two techniques ensures that the reference ranges established for the wide variations of amino/imino acid levels in biological samples with age and sex can be referred to.

Analytical Performance

The overall within and between batch analytical imprecision of the ASTED/HPLC using the total amino acid derivatisation method and utilising the ternary gradient system is shown in Table 6. The other gradient separations performed to a similar specification. The imprecision range depended on the peak shape and fluorescence response and urine samples were more problematical. The mean analytical recovery of the amino/imino acids was 98% (c.v. 3.1%) established from the supplemented samples used in the imprecision studies.

The limit of detection is difficult to assess due to the flexibility of the ASTED system. When combining the amino acid derivatisation with trace enrichment it should be feasible to quantitatively determine some of the amino acids at low fmol concentration on column. Again this will depend on the fluorescence response and chromatographic peak shape. With injection of the dialysates the analytical range was estimated to be 0.5 to 8000 pmol on column, calculated from the recovery of amino acid using dialysis and the kinetics of the OPA/MCE and FMOC reaction. Overall, the analytical response of the method was linear up to 1.5 mmol/l of each amino/imino acid in the original sample.

Chromatographic and Sample Preparation Interferences

The interference of ammonia has already been discussed. However, along with the precipitate of the ammonia-TPB complex, the insoluble excess FMOC and its hydrolysis product are removed by membrane purification. There is always the potential using on-line automation procedures that transfer tubings and the membrane will gradually retain some of this particulate matter. Even with the purge volumes employed, which avoided inter-ferences of sample to sample interaction, retention of particulate matter was observed over a month of continual sampling. Both the ammonia-TPB complex and the excess FMOC are soluble in ketonic organic solvents and treatment of the donor side of the system at weekly intervals with 70% acetone in water negated this interference.

Table 6.   Imprecision of the HPLC/ASTED method.

Sample	Within Batch Imprecision (C.V.%)	Between Batch Imprecision (C.V.%)	Amino Acid Range (μmol/l)
Serum			
Level 1	2.5 - 9.5	3.5 - 12.2	25 - 768
Level 2	2.1 - 4.3	2.8 - 5.8	315 - 968
Urine			
Level 1	1.9 - 10.3	4.5 - 15.1	30 - 987
Level 2	1.8 - 8.6	3.8 - 8.7	382 - 1345

The use of regeneratable membranes to minimise the chromatographic interference of FMOC and its hydrolysis product is shown in Figure 10. The filtration properties of the membrane, and the lower recoveries of static dialysis, was a considerable advantage (espe-cially given the 100 fold increased concentration of FMOC in the dialysis procedure com-pared with direct injection of FMOC). The obvious benefit of this approach is that less re-strictions are placed on the reactant concentrations and biological samples can be derivatised without prior treatment. No chromatographic interferences by such compounds as antibiotics and cephalosporins were observed. Such molecules as penicillamine and histamine did produce a fluorophore with OPA/MCE but, under normal circumstances, their *in vivo* con-centrations would be too low to cause problems.

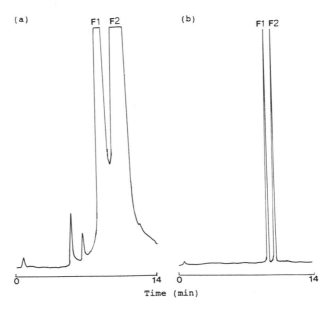

Figure 10. Gradient HPLC of (a) direct injection of 50 μmol/l aqueous FMOC solution; (b) injection of dialysate of 5 mmol/l aqueous FMOC solution using the ASTED procedure described. A rapid gradient program was utilisedidentical to that used for the results obtained in figure 8.

CONCLUSION

The use of HPMP for the preparation of samples prior to amino acid analysis has been shown to provide complete on-line automation. Whilst it is impossible to produce one technique that can accommodate every situation for amino acid analysis, the ASTED approach has been shown to have sufficient flexibility to provide solutions to complex analytical problems. Other derivatisation and analysis strategies can be developed with relative ease. Moreover once an analytical method has been developed the system can be simple to operate.

ACKNOWLEDGEMENTS

Most of the experimental work was conducted in the Biochemistry Department, Coventry and Warwickshire Hospital, Stoney Stanton Road, Coventry. For this I have to thank the staff of that department in particular Mr. R.W. Richardson, Dr S. Jones and Mr M. Lewis, Miss N. Shearsby, Mrs. C. Fuller, Mr. J. McIntosh and many others. I also thank staff at the Childrens Hospital, Birmingham and Salisbury General Hospital for amino acid analysis using ion exchange chromatography.

Thanks also to staff at Gilson Medical Electronics especially Mr E. D'Autry, Dr. F. Verillon, Dr B. Pichon and Mr. F. Qian. Finally it would be remiss of me not to thank Mr. J. Spence at Clinical Innovations Ltd. whose patience permitted the final completion of this report.

REFERENCES

1.   A.J.P. Martin and R.L.M. Synge. A new form of chromatogram employing two liquid phases 1. A theory of chromatography 2. Application to the micro-determination of the higher monoamino-acids in proteins. *Biochem J.*, 35, 1358-1387 (1941).

2.   Z. Deyl, J. Hyanek and M. Horakova. Profiling of amino acids in body fluids and tissues by means of liquid chromatography. *J. Chromatogr.*, 379, 177-250 (1986).

3.   M.D. Armstrong and U. Stave (1973). Plasma free amino acid levels. Variations during growth and age. *Metab. Clin. Exp.* 22, 571-578.

4.   J. Bergstrom, P. Furst, L.O. Noree and Vinnars (1974). Intracellular free amino acid concentrations in human muscle tissue. *J. Appl. Physiol.* 36, 693-697.

5.   L. Hagenfeldt and A. Arvidsson (1980). The distribution of amino acids between plasma and erthyrocytes. *Clin. Chim. Acta.* 150, 133-141.

6.   M. Bordasfonfrede, S. Troupel, G. Petit, J. Glomaud and A. Galli (1985). Study of free amino acid concentration in lacrimal fluid using ion exchange chromatography in the healthy adult. *Ann. Biol. Clin. (Paris)* 43, 387-388.

7.   Y. Houpert, P. Tarallo and G. Siest (1976). Amino acid analysis by ion exchange chromatography using a lithium elution gradient. Influence of methanol concentration and sample pH. *Clin. Chem.* 22, 1618-1631.

8.   W.H. Stein and S. Moore (1954). The free amino acids of human blood plasma. *J. Biol. Chem.* 211, 915-925.

9.   S. Moore, D.H. Spackman and W.H. Stein (1958). Chromatography of amino acids on sulfonated polystyrene resins. *Anal. Chem.* 30, 1185-1205.

10.   M.A. Edwards, S. Grant, and A. Green (1988). A practical approach to the investigation of amino acid disorders. *Ann. Clin. Biochem.* 25, 129-141.

11.   Y. Watanabe and K. Imai (1982). Pre-column labelling for high-performance liquid chromatography of amino acids with 7-fluoro-4-nitrobenzo-1-oxa-1,3-diazole and its application to protein hydrolysates. *J. Chromatogr.* 239, 723-732.

12.   K. Kaneda, M. Sato and K. Yagi (1982). Analysis of dansyl amino acids by reversed-phase high-performance liquid chromatography. *Anal. Biochem.* 127, 49-54.

13.   M.R. Downing and K.G. Mann (1976). High-pressure liquid chromatographic analysis of amino acid phenylthiohydantoins: comparison with other techniques. *Anal. Biochem.* 74, 298-319.

14.   R. Cunico, A.G. Mayer, T. Wehr and T.L. Sheehan (1986). High sensitivity amino acid analysis using a novel automated pre-column derivatisation system. *Biochromatography* 1, 8-14.

15.   S.S. Simons and D.F. Johnson (1978). Reaction of o-phthalaldehyde and thiols with primary amines: formation of 1-alkyl(and aryl)thio-2-alkylisoindoles. *J. Org. Chem.* 43, 2886-2891.

16.   M. Roth (1971). Fluorescence reaction for amino acids. *Anal. Biuochem.* 43, 880-882.

17.   D.C. Turnell and J.D.H. Cooper (1982). Rapid assay for amino acids in serum or urine by pre-column derivatisation and reversed-phase liquid chromatography. *Clin. Chem.* 28, 527-531.

18.   M. Griffin, S.J. Price and T. Palmer (1982). A rapid and sensitive procedure for the quantitative determination of plasma amino acids. *Clin. Chim. Acta.* 125, 89-95.

19.   D.L. Hogan, K.L. Kraemer and J.I. Isenberg (1982). The use of high performance liquid chromatography for quantification of plasma amino acids in man. *Anal. Biochem.* 127, 17-24.

20.   B.N. Jones, S. Paabo and S. Stein (1981). Amino acid analysis and enzymatic sequence determination of peptides by an improved o-phthalaldehyde pre-column labelling procedure. *J. Liq. Chromatogr.* 4, 565-586.

21.   B.N. Jones and J.P. Gilligan (1983). o-phthalaldehyde pre-column derivatisation and reversed-phase high-performance liquid chromatography of polypeptide hydrolysates and physiological fluids. *J. Chromatogr.* 266, 471-482.

22.   D.W. Hill, F.H. Walters, T.D. Wilson and J.D. Stuart (1979). High-performance liquid chromatgraphic determination of amino acids in the picomole range. *Anal. Chem.* 51, 1338-1341.

23.   V.K. Svedas, I.J. Galaev, I.L. Borisov and I.V. Berezin (1980). The interaction of amino acids with o-phthalaldehyde: a kinetic study and spectrophotometric assay of the reaction product. *Anal. Biochem.* 101, 188-195.

24.   S.S. Simons and D.F. Johnson (1977). Ethanethiol: a thiol conveying improved properties to the fluorescent product of o-phthalaldehyde and thiols with amines. *Anal. Biochem.* 82, 250-254.

25.   H. Godel, T. Graser, P. Foedldi, P. Pfaender and P. Fuerst (1984). Measurement of free amino acids in human biological fluid by high-performance liquid chromatography. *J. Chromatogr.* 297, 49-61.

26.   K. Venema, W. Leever, J.O. Bakker, G. Haayer and J. Korf (1983). Automated pre-column derivatisation device to determine neurotransmitter and their amino acids by reversed-phase high-performance liquid chromatography. *J. Chromatogr.* 260, 371-376.

27.   M.J. Winspear and A. Oaks (1983). Automated pre-column amino acid analyses by reversed-phase high-performance liquid chromatography. *J. Chromatogr.* 270, 378-382.

28.   C. Cloete (1984). Automated optimised high-performance liquid chromatographic analysis of pre-column yoy-phthalaldehyde-amino acid derivatives. *J. Liq. Chromatogr.* 7, 1979-1990.

29. D.C. Turnell and J.D.H. Cooper (1983). Automated pre-column derivatisation and its application to amino-acid analysis using high-performance liquid chromatography. *J. Auto. Chem.* 5, 36-39.

30. J.D.H. Cooper, G. Ogden, J. McIntosh and D.C. Turnell (1984). The stability of OPA/mercaptoethanol derivatives of amino acids: an investigation using high-pressure liquid chromatography with a pre-column derivatisation technique. *Anal Biochem.* 142, 98-102.

31. J.D.H. Cooper and D.C. Turnell (1982). Fluorescence detection of cystine by o-phthalaldehyde derivatisation and its separation using high-performance liquid chromatography. *J. Chromatogr.* 227, 158-161.

32. K.S. Lee and D.G. Drescher (1978). Fluorometric amino-acid analysis with ortho-phthalaldehyde (OPA). *Int. J. Biochem.* 9, 457-467.

33. D.C. Turnell and J.D.H. Cooper (1984). Removal of ammonia from urine by tetraphenylboron before amino acid analysis. *Clin. Chem.* 30, 588-589.

34. J.D.H. Cooper, M.T. Lewis and D.C Turnell (1984). Pre-column o-phthalaldehyde derivatisation of amino acids and their separation using reversed-phase high-performance liquid chromatography. I: Detection of the imino acids hydroxyproline and proline. *J. Chromatogr.* 285, 484-489.

35. J.D.H. Cooper, M.T. Lewis and D.C. Turnell (1984). Pre-column o-phthalaldehyde derivatisation of amino acids and their separation using reversed-phase high-performance liquid chromatography. II: Simultaneous determination of amino and imino acids in protein hydrolysates. *J. Chromatogr.* 285, 490-494.

36. L.A. Carpino and G.Y. Han (1970). The 9-fluoro-methoxy-carbonyl function, a new base-sensitive amino-protecting group. *J. Am. Chem. Soc.* 92, 5748-5749.

37. S. Einarsson, B. Josefsson and S. Lagerkvist (1983). Determination of amino acids with 9-fluorenylmethyl chloroformate and reversed-phase high-performance liquid chromatography. *J. Chromatogr.* 282, 609-618.

38. S. Einarsson and B. Josefsson (1987). Separation of amino acid enantiomers and chiral amines using precolumn derivatisation with (+)-1-9-(9-fluorenyl)ethyl chloroformate and reversed phase liquid chromatography. *Anal. Chem.* 59, 1191-1195.

39. S. Einarsson (1985). Selective determination of secondary amino acids using precolumn derivatisation with 9-fluoroenylmethylchloroformate and reversed-phase high-performance liquid chromatography. *J. Chromatogr.* 348, 213-220.

40. B. Gustavsson and I. Betner (1990). Fully automated amino acid analysis for protein and peptide hydrolysates by precolumn derivatisation with 9-fluorenylmethylchloroformate and 1-aminoadamantane. *J. Chromatogr.* 507, 67-77.

41. R. Schuster (1988). Determination of amino acids in biological, pharmaceutical, plant and food samples by automated precolumn derivatisation and high-performance liquid chromatography. *J. Chromatogr.* 431, 271-284.

42. G. Blundell and G. Brydon (1987). High performance liquid chromatography of plasma amino acids using orthophthalaldehyde derivatisation. *Clin. Chim. Acta.* 170, 79-84.

43. D.C. Turnell and J.D.H. Cooper (1985). Automated preparation of biological samples prior to high pressure liquid chromatography. Part I: The use of dialysis for deproteinising serum for amino acid analysis. *J. Autom. Chem.* 7, 177-180.

44. J.D.H. Cooper and D.C. Turnell (1985). Automated preparation of biological samples prior to high pressure liquid chromatography. Part II: The combined use of dialysis and trace enrichment for analysing biological material. *J. Autom. Chem.* 7, 181-184.

45. J.D.H. Cooper and D.C. Turnell (1986). Automated preparation of human samples for analysis of the drug enoximone and its sulphoxide metabolite using high performance liquid chromatography. *J. Chromatogr.* 380, 109-116.

46. D.C. Turnell and J.D.H. Cooper (1987). Automated sequential process for preparing samples for analysis by high-performance liquid chromatography. *J. Chromatogr.* 395, 613-621.

47. D.C. Turnell and J.D.H. Cooper, B. Green, G. Hughes and C.J. Wright (1988). A totally automated HPLC assay for cortisol and cortisone in serum and urine. *Clin. Chem.* 34, 1816-1820.

48. B. Green, J.D.H. Cooper and D.C. Turnell (1989). A fully automated technique for the estimation of urinary free catecholamines. *Ann. Clin. Biochem.* 26, 361-367.

49. M.M.L. Aerts, W.M.J. Beek and U.A.Th. Brinkman (1990). On-line combination of dialysis and column switching liquid chromatography as a fully automated sample preparation technique for biological samples: Determination of nitrofuran residues in edible products. *J. Chromatogr.* 500, 453-468.

50. H. Irth, G.J. De Jong, H. Lingeman and U.A.Th. Brinkman (1990). Liquid chromatograpic determination of azidothymidine in human plasma using on-line dialysis and preconcentration on a silver(I)-thiol stationary phase. *Anal. Chim. Acta.* 236, 165-172.

51. J.D.H. Cooper, D.C. Turnell, B. Green and F. Verillon (1988). Automated sequential trace enrichment of dialysates and robotics. A technique for the preparation of biological samples prior to high-performance liquid chromatography. *J. Chromatogr.* 456, 53-69.

52. J.D.H. Cooper, D.C. Turnell, B. Green, D.J. Wright and E. Coombes (1988). Why the assay of serum cystine by protein precipitation and chromatography should be abandoned. *Anal. Clin. Biochem.* 25, 577-582.

53. B.K. Matuszewski, R.S. Givens, K.Srinivasachar, R.G. Carlson and T Higuchi (1987). N-substituted 1-cyanobenz(f)isoindole: evaluation of fluorescence efficiencies of a new fluorogenic label for primary amines. *Anal. Chem.* 59, 1102-1105.

54. D.C. Turnell and J.D.H Cooper (1989). The automation of liquid chromatographic techniques in biomedical analysis: A critical review. *J. Chromatogr.* 49, 59-83.

55. G. Wittig (1949). Boron-alkali metallo-organic complexes. *Annalhen* 563, 110-115.

56. R.D. McDowall (1989). Sample preparation for biomedical analysis. *J. Chromatogr.* 492, 3-58.

57. R.W. Frei and K. Zech (1988). Selective sample handling and detection in high-performance liquid chromatography. *J. Chromatogr.* 39A.

58. R.W. Frei, M.W.F. Nielen and Th.U.A. Brinkman (1986). Handling of environmental and biological samples via pre-column technologies. *Intern. J. Environ. Anal. Chem.* 25, 3-35.

59. J. Blanchard (1981). Evaluation of the relative efficacy of various techniques for de-proteinising plasma samples prior to high-performance liquid chromatographic analysis. *J. Chromatogr.* 226, 455-460.

60. B.F. Rocks and C. Riley (1986). Automatic analysers in Clinical Chemistry. *Clin. Phys. Physiol. Measure* 7, 1-29.

61. D.C. Turnell and J.D.H. Cooper (1986). A concurrent process for the automatic preparation of biological samples combined with high pressure liquid chromatographic analysis. *J. Autom. Chem* 8, 151-154.

62. F. Flentge, F. Venema and J. Korf (1974). Automated assay of tryptophan in CSF and of total and non-protein bound tryptophan in serum. *Biochem. Med.* 11, 234-241.

63. T. Gerritsen, M.L. Rehberg and H.A. Waisman (1965). Determination of free amino acids in serum. *Anal. Biochem.* 11, 460-466.

64. E. Adams, S. Ramaswamy and N. Lamon (1978). 3-hydroxyproline content of normal urine. *J. Clin. Invest.* 56, 1482-1487.

65. J.M. Rattenbury and J.C. Townsend (1990). Establishment of an external quality assessment scheme for amino acid analyses: results from assays of samples distributed during two years. *Clin. Chem.* 36, 217-224.

66. Geigy Scientific Tables (1970). Letner C. Eds. Ciba-Geigy, Basle.

INTRODUCTION TO AUTOMATED SEQUENTIAL TRACE ENRICHMENT OF DIALYSATES (ASTED) AND ITS APPLICATION TO THE ANALYSIS OF NUCLEOSIDES IN PLASMA

A.R. Buick and C.T.C. Fook Sheung

Department of Bioanalytical Sciences
Wellcome Research Laboratories
Beckenham, Kent, BR3 3BS. UK

SUMMARY

The principles of operation and optimisation of the Gibson ASTED system are described. Examples are given of its application to the analysis of nucleosides in plasma.

INTRODUCTION

The determination of concentrations of drugs in biological samples is an essential part of the development and registration of new pharmaceutical products. Although there are many rate limiting steps along the road leading to a successful marketable medicine, a delay in producing an analytical method to support a study is often less well tolerated than other stages of development. The bioanalyst must therefore plan carefully to give the maximum chance of early success.

There is a choice of sample preparation techniques such as liquid/liquid and liquid/solid phase extraction, protein precipitation, ultrafiltration and column switching [1] and a choice of separation methods such as gas chromatography (GC), high performance liquid chromatography (HPLC) and thin-layer chromatography (TLC). Increasingly techniques are being combined to produce on-line automated sample clean-up and quantification with which to improve efficiency. However, automation does not remove from the bioanalyst the decision making process involving selection of analytical parameters to suit the physical and chemical characteristics of the drug.

The ASTED (Automated Sequential Trace Enrichment of Dialysates) automates the combination of dialysis and column switching sample preparation techniques [2] and is most often coupled to HPLC with UV detection.

Sample Preparation for Biomedical and Environmental Analysis,
Edited by D. Stevenson and I.D. Wilson, Plenum Press, New York, 1994

Plasma contains high concentrations of complex macromolecules such as proteins, carbohydrates and lipids. They must be separated from the low molecular weight drugs for analysis to prevent damaging the HPLC column and to reduce the number of compounds which would interfere with quantification. Although dialysis is a good candidate for achieving separation of low and high molecular weight substances, it suffers from two serious deficiencies as far as on-line sample preparation is concerned. Firstly, the diffusion process of transfer of compounds across the dialysis membrane is slow. Secondly, the amount of compound recovered in the dialysate is low and in a much diluted solution. Analysis of a portion of such a diluted solution using HPLC would result in considerably worsened limits of quantification.

The parameters controlling movement of molecules across a membrane during dialysis are described by Fick's law as shown below:

$$\frac{dm}{dt} \quad -DA \quad \frac{dc}{dx}$$

where

D	=	diffusion coefficient	$\frac{dm}{dt}$	=	mass flow diffusion rate
A	=	membrane surface area	$\frac{dc}{dx}$	=	concentration gradient

It can be seen that once the system has been defined in terms of the matrix to be dialysed (donor solution), liquid to receive the dialysed compound (recipient solution) and the geometry of the dialysis cell (membrane surface area) then the concentration gradient is the controlling force for transfer of molecules to the dialysate. The greater the concentration difference between donor and recipient solutions, the greater will be the rate of dialysis. In the ASTED system developed by Cooper & Turnell [3] an increased concentration gradient is achieved by constantly replacing the recipient solution during dialysis with fresh solution in a flowing system.

It is obvious that although the quantity of analyte recovered is increased the resulting solution is diluted to such a degree as to make it unusable for analysis without further processing. Concentration of the much diluted solution is achieved by passing it through a cartridge of HPLC packing material to retain the drug while allowing the liquid to drain to waste. In addition to trace enrichment of the drug, continued passage of recipient solution through the cartridge, provided retention of analyte(s) is maintained, causes improved cleanup by increased elution of endogenous components. The choice of cartridge packing material, solutions etc., must, of course, be made by the chromatographer as for any chromatographic system.

After enrichment has taken place the cartridge is switched in-line with the analytical column so that HPLC mobile phase elutes the drug for separation and quantification according to previously determined conditions. During the separation and quantification step the cartridge is regenerated by washing with recipient solution. The process is then repeated in a continuous and automatic manner.

EXPERIMENTAL

Equipment

Gilston ASTED and HPLC system with UV detection.

HPLC Conditions

Analytical column, 150 x 4.6mm Spherisorb 5-ODS-2
Enrichment cartridge, Prelute ODS (Anachem)
Mobile phase, 5% acetonitrile in 4mmol/l aqueous ammonium acetate
Flow rats, 1.5ml/min
Detection wavelength, 290nm

Sample

Plasma centrifuged and loaded directly into vials in the ASTED autosampler rack.
Protein unbinding reagent, 0.5mol/l aqueous monochloracetic acid automatically added prior
to dialysis.

Typical Performance

Limit of quantification, 0.2μmol/l, approximately 50ng/ml.
Accuracy and precision, see tables 1 & 2.

Dialysis

Dialysis time typically 3 or 5 min recipient phase flowing.
Enrichment volumes typically 600-1200 μl.

RESULTS

Optimisation Of ASTED Operating Conditions

Clearly two important variables for optimising ASTED conditions are the volume of
solution flowing through the cartridge during concentration of the dialysate and the volume
allowable before unacceptable elution of analyte to waste occurs. It may be imagined that an
infinite amount of fresh, continuously flowing recipient solution would eventually provide
close to quantitative recovery of analyte onto the cartridge as illustrated in Figure 1a. In
practice, leaching of the compound occurs well before complete recovery and an
enrichment/elution profile is produced of the type illustrated in Figure 1b. The optimum
volume of recipient solution to be used therefore is defined by that at, or around, the maximum of the profile. The maximum is typically represented by a volume of 1 to 2ml of solution passing through the trace enrichment cartridge in approximately 3 to 5 minutes. The
profile is often referred to as a breakthrough curve. Each compound has its own characteristic breakthrough curve for the conditions used and once determined allows analysis to proceed with the best recovery achievable.

The ASTED equipment consists of a versatile autosampler to pass aliquots of samples directly, or after programmed mixing with reagent, to the donor channel of the dialysis
block in addition to column switching equipment necessary for enrichment and subsequent
elution of the analyte onto the analytical column, Figure 2. Control is by a computer which

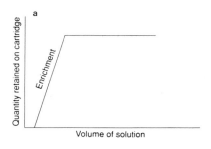

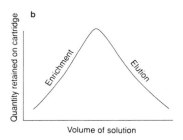

Figure 1a and b. "Ideal enrichment profile" and a more typical enrichment/elution profile of the type encoun
tered in practice.

can be programmed to carry out many analyses of standard solutions to enable a breakthrough curve to be determined overnight.

It is not mandatory to select conditions represented by the maximum of the breakthrough curve. Often, cleaner chromatograms can be obtained by selecting a volume of
recipient solution greater or less than optimum for recovery where maximum sensitivity is
not required. The situation frequently occurs where concentrations of more than one
compound must be determined as in the case of a drug and its metabolite. Since each
compound will have its own characteristic breakthrough it is unlikely that the best conditions
will be the same for both. This situation is illustrated diagramatically in Figure 3. A choice
therefore, has to be made since the recovery of one or both may suffer from the parameters
selected. Using best conditions for one will give a better limit of quantification than for the
other and vice versa.

The analyst may well have other factors to consider to assist with the decision.
Optimum conditions for one compound may give a cleaner chromatogram for the other
allowing sensitivity to be increased. One of the components may have a much better/nar-
rower peak or higher extinction coefficient, fluorescence etc. allowing optimum conditions
for the other to be used. An example of two chromatograms resulting from analysis of the 5-
propynylarabinofuranosyl analogue of deoxyuridine (PYaraU) and metabolite is shown in
Figure 4 at the two optima. In this instance it can be seen that the metabolite peak changes
only slightly between chromatograms while a large change is seen in the size of the PYaraU
peak.

Since both chromatograms are clean the optimum conditions for PYaraU can be used.
This is a good illustration of the fact that although predictions of choice of conditions are
useful, only when the actual chromatograms and breakthrough curves are produced can
decisions be made for particular compounds.

Protein Binding

Interactions of the drug with the matrix of a biological sample may require other

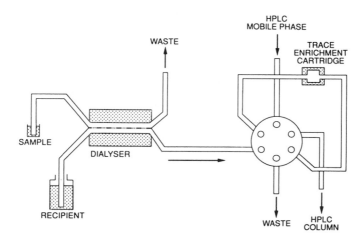

Figure 2. Schematic diagram of the ASTED system.

factors to be considered in automated sample preparation. Drugs are frequently bound to proteins which prevents them from passing through the dialysis membrane with the consequence that low recoveries are obtained. Although a great advantage of the ASTED equipment is the high reproducibility of data even at absolute recoveries which may be as low as 1-2%, to achieve an acceptable limit of quantification the drug must be unbound to present the maximum amount of free drug for dialysis. Off-line protein precipitation is an obvious suggestion to solve the problem but one would have to consider the physical effect of any precipitating solvents on the dialysis block materials, the dilution effect of such additions and the increased operator time. Other sample preparation techniques for deproteinisation such as ultrafiltration are not without their problems [4].

Some drugs can be released from proteins by adding a reagent to compete for binding. As described later this presents an easily automated mechanism to improve recovery during dialysis.

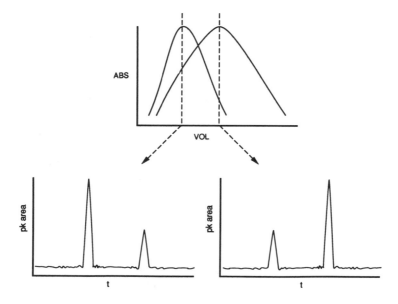

Figure 3. A systematic representation of breakthrough curves for two compounds.

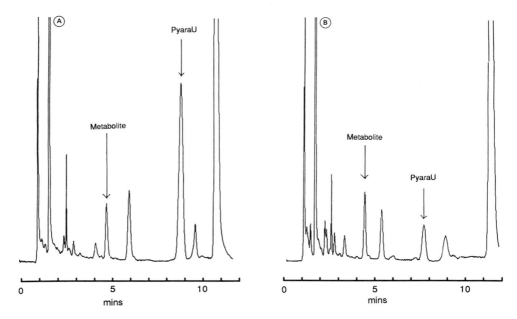

Figure 4.   Chromatograms representing choice of optimum conditions for PYaraU and its metabolite. Chromatographic conditions were as in experimental section.

Determination Of Nucleosides In Plasma

Analysis of nucleoside drugs in biological matrices presents a special challenge because of their similarity to naturally occurring substances. Their thermal instability, low volatility and high polarity give HPLC something of an advantage over other methods such as GC in which an additional sample preparation stage of derivatisation may be required.

The ASTED system has been successfully applied to the analysis of nucleosides. As an example a selection of common nucleosides were added to plasma and subjected to analysis by ASTED as shown in Figure 5. Breakthrough curves were not determined for each compound; only a typical value suggested from experience was chosen. Even without attempts at optimisation of conditions good separation from endogenous components and between nucleosides and bases was achieved. Improvements are obviously possible for development of a method focusing on a single nucleoside or base.

A nucleoside currently being studied as an antiviral agent is the 5-propynyl arabino-furanosyl analogue of deoxyuridine, PYaraU [5]. An analytical method for the determination of PYaraU and a metabolite,M, in plasma has been developed using ASTED with HPLC (publication in preparation). The method is outlined below.

The enrichment cartridge and analytical column both contained C18 bonded silica gel and the HPLC mobile phase, run isocratically, was a mixture of acetonitrile and ammonium acetate solution. Detection was by UV absorption. Further details are included in the experimental section.

Getting the balance right for concentration and elution from the enrichment cartridge is a critical feature of retention of nucleosides on C18 bonded silica gel. The high polarity of these compounds allows only a small volume of dialysate to pass through the cartridge before a significant amount of elution takes place. A breakthrough curve must therefore be determined carefully to give the maximum volume of enrichment and hence washing, to remove as many interfering and column poisoning components as possible.

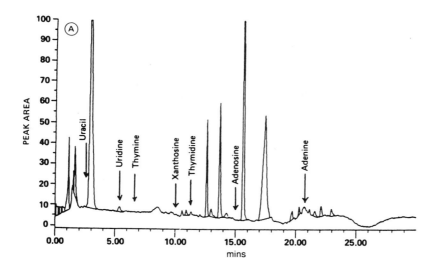

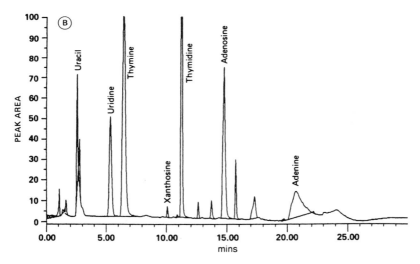

Figure 5. Illustration of the anlaysis of common nucleosides and bases in human plasma by ASTED/HPLC. Chromatographic conditions were as in experimental section.

PYaraU, as with other nucleosides, is extensively bound to plasma protein. Although determination of concentrations of PYaraU and its metabolite gave excellent reproducibility of results the absolute recovery was extremely low (approximately 1 to 2%). Introducing a low concentration of chloracetic acid into the plasma prior to dialysis competed effectively for protein binding and relative recoveries (comparing spiked plasma with aqueous solution) of 100% were then achieved.

With careful adjustment of the proportions of modifier and ammonium acetate solution in the mobile phase acceptably clean chromatograms were obtained. A typical example is shown in Figure 6.

Tables 1 and 2 show precision and accuracy achieved for PYaraU and its metabolite in plasma exemplifying the high reproducibility of data attainable even for such problematical compounds as nucleosides.

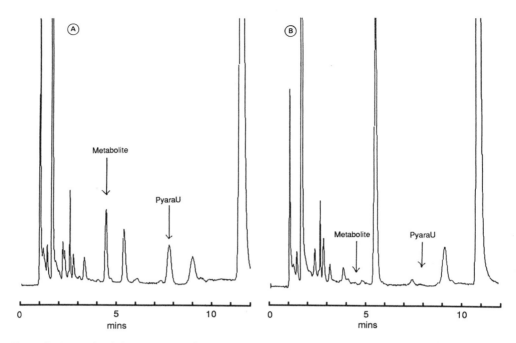

Figure 6. Example of chromatograms of PYaraU and metabolite in (a) spiked and (b) drug-free human plasma. Chromatographic conditions were as in the experimental section.

It is accepted that the method is capable of further optimisation and improvement but it has been used to routinely produce many thousands of results and has therefore proved its worth. Alternative methods such as solid phase extraction failed to produce adequate selectivity and sensitivity.

CONCLUSION

The ASTED in combination with HPLC is a rapidly expanding automated sample preparation and analysis technique. Although demanding careful consideration of parameters with which to develop bioanalytical methods the ASTED provides an efficient and accurate alternative to other widely used techniques.

Table 1. Typical intra-assay data

Concentration μmol/l	PYaraU		Metabolite	
	Accuracy %	*CV%*	*Accuracy %*	*CV%*
0.2	100.8	1.2	101.7	2.1
0.4	99.7	1.5	99.3	1.6
1.0	99.2	4.2	101.7	1.4
5.0	98.2	1.1	101.0	0.7
10.0	102.3	1.0	100.9	1.0
25.0	100.8	0.7	100.3	0.8

n = 6

124

Table 2.   Typical inter-assay data

Concentration µmol/l	PYaraU Accuracy %	CV%	Metabolite Accuracy %	CV%
0.2	105.1	4.1	105.9	3.5
0.4	100.9	3.3	101.1	2.2
1.0	98.9	2.0	100.7	1.3
5.0	100.9	1.6	100.7	1.3
10.0	102.7	1.6	101.0	1.4
25.0	101.5	1.5	100.5	1.6

n = 5

REFERENCES

1.   R.D. McDowall. Sample preparation for biomedical analysis. *J. Chromatogr.* 492, 3-58, 1989.
2.   U.R. Tjaden, E.A. De Bruijn, J. Van der Greet and H. Lingeman. Drug monitoring by on-line dialysis and liquid chromatography. *LC-GC* 5, 32-42, 1988.
3.   J.D.H. Cooper, D.C. Turnell, B. Green and F. Verillon. Automated sequential trace enrichment of dialysates and robotics. *J. Chromatogr.* 456, 53-69, 1988.
4.   D.C. Turnell and J.D.H. Cooper. Automation of liquid chromatographic techniques for biomedical analysis. *J. Chromatogr.* 492, 59-83, 1989.
5.   S.F. Lacey, T. Suzutani, K.L. Powell, D.J.M. Purifoy and R.W. Honess. Analysis of mutations in the thymidine kinase genes of drug resistant varicella-zoster virus populations using the polymerase chain reaction. *Journal of general virology*, 72, 623-630. 1991.

HIGH RESOLUTION ^{1}H NMR SPECTROSCOPIC MONITORING OF EXTRACTION OF MODEL GLUCURONIDES ON PHENYLBORONIC ACID AND C_{18} BONDED PHASES

M. Tugnait[1], I.D. Wilson[2] and J.K. Nicholson[1]

[1] Department of Chemistry
Birkbeck College
Gordon House
29 Gordon Square
London
WC1H 0PP, UK

[2] Department of Safety of Medicines
ICI Pharmaceuticals Division
Mereside
Alderley Park
Macclesfield
Cheshire, SK10 4TG, UK

SUMMARY

Solid-phase extraction chromatography (SPEC) can be used in conjunction with NMR spectroscopy (SPEC-NMR) for the fractionation, isolation, purification, detection, identification and quantification of endogenous substances and xenobiotics from biological fluids. To date the bulk of the work has employed C_{18} bonded silica gel or similar materials with ion-suppression to retain ionisable metabolites followed by step-wise elution and NMR detection. More recently we have begun to investigate the use of other, potentially more specific phases, such as immobilised phenylboronic acid (PBA) which has previously been used for the selective extraction of catecholamines. Here, we report the application of SPEC-NMR to extract phenolphthalein glucuronide and other model drug glucuronides from urine using solid-phase extraction onto disposable columns containing immobilised phenylboronic acid. These results were compared to those obtained for C_{18} bonded silica gel columns using simple step gradients. The results of these preliminary experiments revealed the selectivity and extraction potential of PBA columns for glucuronides and illustrated the differences between the PBA and C_{18} columns.

INTRODUCTION

High resolution ^{1}H nuclear magnetic resonance (NMR) spectroscopy is a rapid and effective means of detecting, identifying and quantifying many drugs, drug metabolites and endogenous compounds in biofluids with minimal pre-treatment of the sample [1]. However, the effectiveness of drug metabolite detection depends on the concentration of the analyte(s) (ca 10 nM/ml minimum), the number of metabolites present and the chemical shift range over which metabolite resonances occur and hence the degree of interference due to

Sample Preparation for Biomedical and Environmental Analysis,
Edited by D. Stevenson and I.D. Wilson, Plenum Press, New York, 1994

127

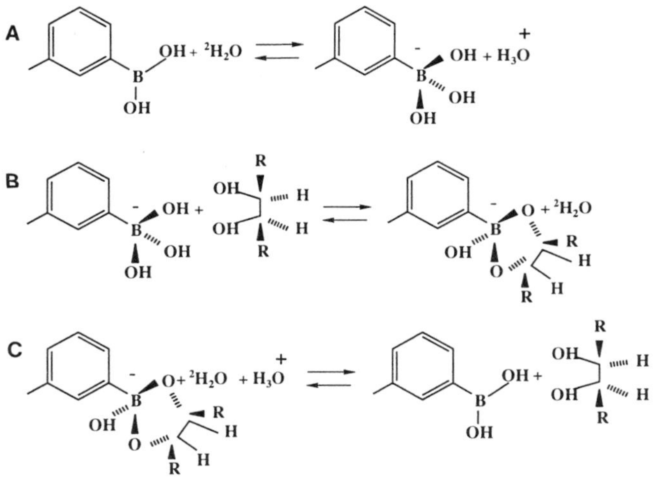

Figure 1. (A) Equilibration with an alkaline solution to give the reactive boronate form $RB(OH)_3^-$. (B) The analyte becomes covalently bound with the concomitant release of H_2O. Once the compound is retained contaminants can be selectively washed from the bonded phase provided an alkaline pH is maintained. (C) Elution by acidification of the boronate complex.

peak overlap from endogenous metabolites. Where peak overlap problems are severe solid-phase extraction chromatography (SPEC) has been shown to be effective for metabolite separation, partial purification and concentration prior to NMR analysis [2] (see also Nicholson and Wilson, this volume). SPE provides a rapid pre-treatment which, in favourable circumstances, can separate the bulk of the endogenous components from drug metabolites in biofluids [3]. Previously we have used C_{18} bonded phase columns extensively to isolate and purify drug glucuronides from complex biofluids, such as urine and bile prior to their further characterisation by mass spectrometry or NMR [2-4]. Recently, we reported our preliminary investigations on the use of potentially more specific phase, immobilised phenylboronic acid (PBA) to extract phenolphthalein glucuronide from urine [5]. The usefulness of boronic acid derivatives attached to insoluble supports for the chromatography of sugars, nucleosides, nucleotides and nucleic acid polymers was first exploited by Weith et al and Rosenberg et al [6,7]. The retention of suitable compounds on PBA (e.g. catecholamines, carbohydrates and vicinal diols) is facilitated by both covalent and non-specific retention mechanisms (Figure 1). The bonding energetics of the covalent retention mechanism are high compared to the non-polar interactions involved in retention by reversed-phase mechanisms. Thus the covalently-bound analyte remains bound to the column whilst compounds retained by non-specific mechanisms can be eluted. PBA columns thus offer a potentially more selective and efficient means of isolating glucuronides from biofluids than C_{18} columns.

In order to test this, we have applied phenylboronate SPEC-NMR to the extraction of a number of model glucuronides spiked into urine and compared the extraction/purification efficiency with those of C_{18} SPEC-NMR methods.

EXPERIMENTAL SECTION

The compounds used in this study were obtained from Sigma UK, and included phenolphthalein glucuronide (PHG), p-nitrophenyl-β-D-glucuronide (PNPG), α-naphthyl-β-D-glucuronide (NG), 6-bromo-2-naphthyl-β-D-glucuronide (BNDG) and phenolphthalein disulphate (PHDS).

128

Sample Preparation

Control urine from a clinically normal volunteer was collected into polypropylene tubes and stored until required. Urine samples spiked with the individual glucuronides were prepared with a final concentration of 5mM. 500µl of each sample was freeze-dried, reconstituted in 600µl of ^2H_2O and placed in 5mm (o.d.) glass NMR tubes containing 50µl (2.5mg/ml) sodium-3-trimethylsilyl-(2,2,3,3,-$^2H_4$)-1-propionate (TSP; δ = 0ppm, as an internal reference) in ^2H_2O.

Solid-Phase Extraction Onto C_{18} Columns

Disposable polypropylene cartridges (3ml Bond-Elut^{TM}) packed with 200mg of C_{18} bonded silica (Varian Associates, obtained from Jones Chromatography Ltd, UK) were used to extract glucuronides from urine, except for PNPG where a 500mg sorbent capacity column was used. The C_{18} extraction and elution protocol was as follows: 500µl of the acidified urine (previously spiked with a model glucuronide) (pH2, using 2M HCl) was loaded onto the activated C_{18} column (5ml of methanol followed by 5ml of acidified water (pH2). Acidification of the sample ensured the retention of the glucuronides on the column. In order to obtain the selective recovery of the glucuronide step-wise elution procedures were employed. The retained compounds were eluted with 1ml of methanol:acidified water mixtures of increasing elutropic strength, i.e. 20:80, 40:60, 60:40, 80:20 and 100% methanol, into scintillation vials. Methanol was removed using nitrogen and water by freeze-drying prior to reconstitution in ^2H_2O (600µl) for spectroscopy.

Solid-Phase Extraction Onto PBA Columns

The PBA protocol for SPE of glucuronides was as follows: 500µl of the spiked urine sample, mixed with 1.5ml of 100mM glycine buffer (pH 8.5), was loaded onto the PBA column (Varian Associates, obtained from Jones Chromatography Ltd, UK). The column was previously equilibrated with 5ml of 100mM glycine buffer (pH 10) followed by 5ml of glycine buffer (pH 8.5) in order to obtain the reactive boronate from $RB(OH)_3$. The retained glucuronide was then eluted with 5ml of methanol-1% HCl (90:10 v/v).

Methanol was removed using nitrogen followed by freeze-drying prior to reconstitution (600µl) for spectroscopy.

NMR Spectroscopy

^1H NMR spectra were recorded on a JEOL GSX500 MHz Spectrometer operating at a field strength of 11.75 T (500 MHz ^1H frequency) over a 6000 Hz spectral width. Typically, 64 free induction decays (FID's) were collected for each sample into 32 K data points with a data accquisition time of 2.73 s. A pulse delay of 2.27 s was used to ensure complete T_1 relaxation of the spectra. An exponential apodisation function corresponding to a line broadening factor of 0.2 Hz was applied prior to Fourier transformation. A secondary gated irradiation field at the water resonance frequency was applied in order to suppress the intense signal from water. Chemical shifts were referenced internally to the singlet resonance of TSP, δ = 0ppm (50µl of a 2.5mg/ml solution was added to each sample).

RESULTS AND DISCUSSION

The 500 MHz ^1H NMR spectrum of freeze-dried control urine is shown in Figure 2A. The spectrum contains several chemical shift "windows" in which there are relatively

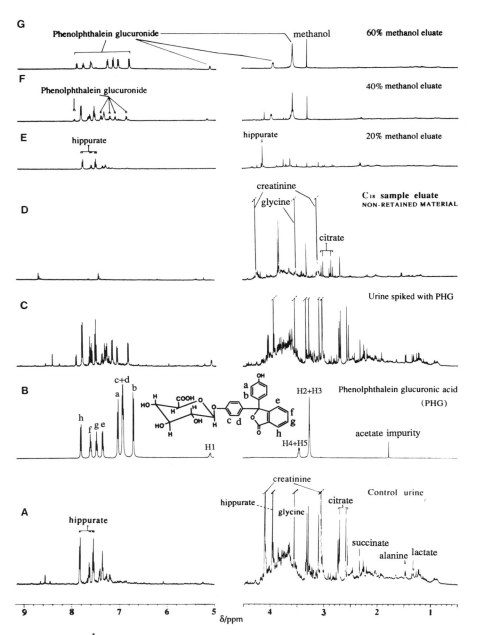

Figure 2. 500 MHz ^{1}H NMR spectra of human urine containing phenolphthalein glucuronide (PHG) after solid-phase extraction on a C_{18} Bond-Elut^{TM} column. (A) control human urine, (B) (PHG in ^2H_2O), (C) urine spiked with PHG, (D) C_{18} non-retained material, (E) 20% methanol eluate, (F) 40% methanol eluate, (G) 60% methanol eluate.

few resonances from endogenous compounds. These ^{1}H NMR "windows", in which signals from the glucuronides may be observed with little interference, cover the chemical shift ranges δ = 0 to 1 ppm, 1.5 to 1.9 ppm, 5 to 7 ppm and 7.8 to 12 ppm. Extraction of an aliquot of acidified control urine resulted in the retention of the bulk of the urinary aromatic acids (e.g. hippurate) whereas most of the non-aromatic amino acids and basic compounds (e.g. creatinine) passed unretained throught the column and were detected in the eluate. Subsequent step-wise elution showed that 20% aqueous methanol (pH 2) was sufficiently elutropic to provide complete recovery of these components. Conversely extraction of control urine samples, mixed with glycine buffer at pH 8.5, on the PBA phase showed little or no retention of endogenous metabolites (e.g. hippurate, citrate, lactate and creatinine). SPE on PBA and C_{18} bonded phases was then applied to urine samples spiked with a variety of model glucuronides (as described below).

Phenolphthalein Glucuronide (PHG)

Initial experiments using PBA investigated the extraction of PHG from urine and the results obtained were compared to those of C_{18} bonded phase extraction [5]. Addition of PHG to urine resulted in several new resonances being observed (Figure 2B and 2C). These included a series of overlapping aromatic resonances between δ = 6.6 to 8.0 ppm, and a prominent doublet at 5.11 ppm from the β-anomeric proton of the glucuronic acid moiety. The spectral region from 3.0 to 3.6ppm also contained several signals from PHGA. SPE of 500μl of the acidified spiked urine sample onto a C_{18} bonded phase cartridge followed by step-wise gradient elution using increasingly elutropic methanol-acidified water mixtures (pH 2) resulted in the series of NMR spectra shown in Figure 2D-G. Removal of the non-retained material (Figure 2D) followed by an acidified water wash and then a 20% aqueous methanol wash (Figure 2E) led to the elution of several endogenous compounds including creatinine, citrate, alanine, lactate and hippurate. Subsequent elution with 40% aqueous methanol resulted in the recovery of 54.4% of the spiked PHG while further elution with 60% aqueous methanol led to a further 34.5% recovery of the glucuronide (Figures 2F and 2G). Small quantities (4.7%) of PHGA were also recovered in the 80% aqueous methanol eluate (not shown). The total recovery in all three fractions was 93.6%. Clearly, having removed the contaminated co-extracted endogenous substances with the 20% methanol wash, virtually complete elution and recovery of the glucuronide could have been accomplished with a subsequent 80% methanol wash.

Extraction of PHG on the PBA-bonded phase was subsequently attempted. Covalent bonding to the PBA is a function of the pH (the pKa of the immobilised phenylboronic acid being approximately 9.2). It was therefore essential to fully equilibrate the bonded phase with an alkaline solution to obtain good extraction efficiency for the PHG. In our initial experiments, we experienced difficulty in retaining all of the PHG on the column, with losses during the inital elution to remove endogenous substances (i.e. column overlead was occuring). Further experiments were performed to determine the mass of the PBA sorbent required to facilitate complete retention of the glucuronide. Four PBA columns containing increasing amounts of sorbent were prepared (i.e. 100, 200, 300 and 400 mg) and the extraction efficiency monitored by NMR. The elution profiles obtained revealed that the 300mg sorbent capacity PBA cartridge retained the bulk of the PHG. All of the PHG was effectively extracted onto the 400mg sorbent capacity PBA cartridge (Figure 3 and Table 1). Essentially, all of the endogenous urinary metabolites such as citrate, lactate, hippurate and creatinine were non-retained by the PBA (Figure 3B). The retained PHG was recovered in a single step by elution with 5ml methanol-1% HCl (90:10 v/v). This protocol gave a 96.7% recovery assigned (Figure 3C). Resonances related to the glycine buffer and to small amounts of creatinine were also present in the spectrum.

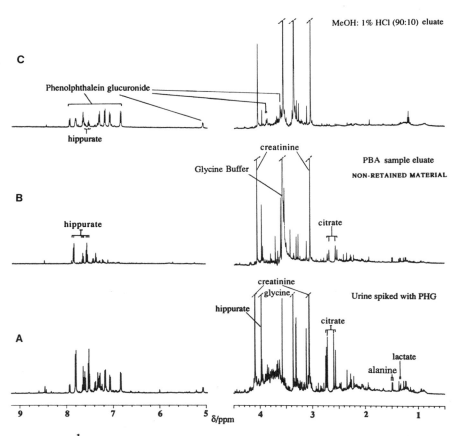

Figure 3. 500 MHz ^{1}H NMR spectra of human urine containing phenolphthalein glucuronide (PHG) after solid-phase extraction on a PBA column. (A) urine spiked with PHG, (B) PBA non-retained material, (C) methanol-1% HCl (90:10) eluate.

p-Nitrophyl-β-D-glucuronide (PNPG)

Solid-phase extraction of PNPG spiked into urine on a C_{18} bonded silica cartridge followed by subsequent elution with a step-wise gradient was monitored by ^{1}H NMR. As with PHG, initial elution of the acidified spiked urine followed by elution with acidified water (pH2) and then 20% aqueous methanol led to the removal of the bulk of the endogenous components. Some 54.1% of the PNPG was also eluted in the 20% aqueous methanol wash. Subsequent elution with 40% aqueous methanol led to a further 36.0% recovery of the glucuronide and 3.3% of the glucuronide was recovered in the 80% aqueous methanol eluate giving an overall recovery of 93.4% of the applied PNPG.

Experiments to extract PNPG (5mM final concentration in urine), onto a 500mg sorbent capacity PNA column were not promising as less than 20% of PHPG was retained on the PBA phase. Columns of larger sorbent capacity (1000mg) were also investigated but total retention of the glucuronide could not be achieved. The alternative approach of extracting smaller amounts of PNPG (2nM final concentration in urine) on both 500mg and 1000mg sorbent capacity columns as investigated but as equally unsuccessful. The effect of pH on the extraction/retention capacity of the column was also examined. Variation of loading pH (i.e. loading sample at pH 9 and pH 10) did not improve the retention of PNPG on the column. These results are summarised in Table 1.

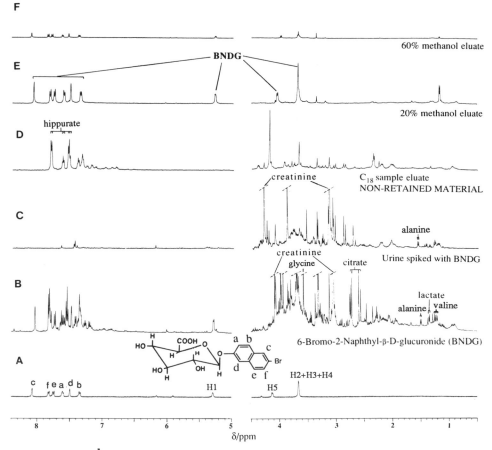

Figure 4.   500 MHz ^{1}H NMR spectra of human urine containing 6-bromo-2-naphthyl-β-D-glucuronide (BNDG) after solid-phase extraction on a C_{18} Bond-Elut^{TM} column. (A) BNDG in ^2H_2O), (B) urine spiked with BNDG, (C) C_{18} non-retained material, (D) 20% methanol eluate, (E) 60% methanol eluate, (F) 80% methanol eluate.

α-Naphthyl-β-D-glucuronide (NG)

Concentration of NG from urine onto a C_{18} column followed by elution with acidified water (pH 2) and then 20% aqueous methanol resulted in the removal of endogenous compounds with small amounts of the glucuronide (0.3%) appearing in the acidified water wash. Step-wise elution with a gradient of increasing elutropic strength, allowed the complete removal of most endogenous components, and recovery of the glucuronide in three fractions. Elution with 40% aqueous methanol resulted in a 38.6% recovery of the glucuronide whilst a further recovery of 40.1% was obtained in the 60% aqueous methanol wash. 5.5% of the glucuronide was recovered in the 80% aqueous methanol eluate. The overall recovery in the three fractions was 84.5%. As with PHGA it is evident that, following the removal of the bulk of the endogenous metabolites with 20% aqueous methanol, all of the NG could have been recovered in a single 80% methanol wash.

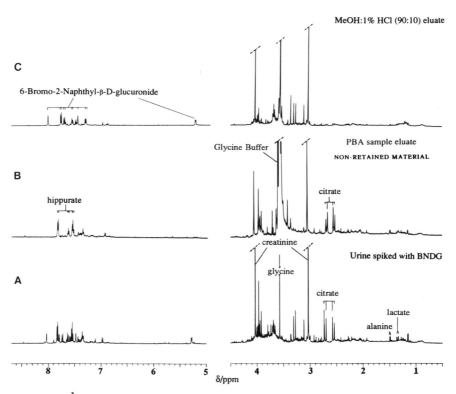

Figure 5. 500 MHz ^{1}H NMR spectra of human urine containing 6-bromo-2-naphthyl-β-D-glucuronide (BNDG) after solid-phase extraction on a PBA column. (A) urine spiked with BNDG, (B) PBA non-retained material, (C) methanol-1% HCl (90:10) eluate.

Extraction of NG from onto a 500mg sorbent capacity PNA column revealed similar problems to those experienced with PNPG, with considerable breakthrough of the glucuronide. Initial elution of the spiked urine resulted in the co-elution of 48.3% of the glucuronide with the non-retained endogenous urinary metabolites. The retained glucuronide was eluted (51.7%) with methanol-1% HCl (90:10 v/v). The use of larger sorbent capacity columns and the extraction of smaller amounts of the glucuronide (1mM final concentration of spike in urine) on a 500mg sorbent capacity column were investigated but complete retention of the glucuronide could not be obtained (see table).

6-Bromo-2-naphthyl-β-D-glucuronide (BNDG)

^{1}H NMR spectra obtained from urine spiked with BNDG contained readily discernable resonances due to the glucuronide (Figure 4B) with a small doublet at 5.29ppm corresponding to the β-anomeric proton of the glucuronide ring. Extraction of this urine sample onto a C_{18} column followed by subsequent elution with acidified water resulted in the removal of the bulk of the endogenous compounds. When processed by using step-wise gradients the glucuronide was recovered in three fractions essentially free of endogenous compounds. Thus the 40% aqueous methanol fraction contained 2.0% of the glucuronide (not shown) wherease the bulk of the BNDG (82.4%) was recovered in the 60% aqueous

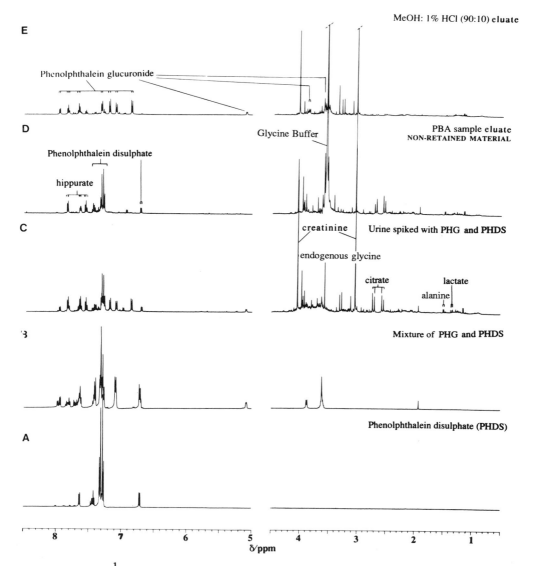

Figure 6. 500 MHz ^{1}H NMR spectra of human urine containing a mixture of PHG and phenolphthalein disulphate (PHDS) after solid-phase extraction on a PBA column. (A) PHDS in ^{2}H$_2$O, (B) mixture of PHG and PHDS (pH in ^{2}H$_2$O = 9.50), (C) urine spiked with PHG and PHD, (D) PBA non-retained material, (E) methanol-1% HCl (90:10) eluate.

methanol eluate (Figure 4E). Further elution with 80% aqueous methanol (Figure 4F) recovered an additional 9.8% of the glucuronide. A total recovery of 94.2% was achieved on the C_{18} bonded silica gel column.

Extraction of BNDG was then investigated on a 600mg sorbent capacity PBA column (as slight breakthrough of the glucuronide was observed on a 500mg capacity column). Inital elution of the spiked urine with glycine buffer at pH 8.5 led to the removal of interfering metabolites with complete retention of the glucuronide on the column (Figure 5B). The retained glucuronide was recovered in a single step using 5ml of methanol-1%

HCl (90:10 v/v) essentially free of endogenous metabolites (Figure 5C). A >90% recovery of the glucuronide was obtained.

The results of the extraction on the PBA phase for all the model glucuronides are summarised in Table 1.

Use Of PBA Columns To Concentrate Dilute Solutions Of Glucuronides

In addition to the use of SPEC-NMR to fractionate complex mixtures (2-4), it can also be used to provide a concentrate of analytes from urine. We therefore attempted the extraction of 5ml of a 1mM solution of PHG in urine onto a PBA column with the aim of concentrating small amounts of the glucuronide from relatively large volumes of urine. This study resulted in the recovery of >90% of the glucuronide demonstrating that, for suitable compounds, extraction and concentration onto the PBA phase can be successful accomplished.

Fractionation Of Sulphates And Glucuronides (PHG and PHDS)

Metabolism of phenolic drugs often results in the production of a mixture of sulphate and glucuronide conjugates. The extraction and separation of PHGA and PHDS (Figure 6A) spiked into control urine was attempted. The final concentration of each compound in urine was 5mM (Figure 6C). Solid-phase extraction was performed on a 500mg PBA column and the results obtained were as follows: Initial elution of the spiked urine sample mixed with 1.5ml of glycine buffer at pH 8.5 led to the removal of endogenous metabolites together with all of the PHDS which was completely unretained (Figure 6D). Elution with 5ml methanol-1% HCl (90:10 v/v) recovered PHG (Figure 6E) essentially free of endogenous metabolites and PHDS (although glycine from the buffer was present).

The results of these studies clearly highlight the advantages and limitations of the PBA phase as a means of selectivity extracting and purifying glucuronides. Thus, the high specificity of PBA for PHG enabled the selective extraction, concentration and purification of this metabolite from both endogenous compounds and the related disulphate. However, it is also apparent from these results that PBA does not provide a general method for the extraction of glucuronides from biofluids, as evidenced by the disappointing results obtained for PNPG and NG. This would suggest that some aspects of the extraction of these glucuronides onto PBA are controlled by secondary interactions of one sort or another. The nature of these interactions is unclear, but it is possible they involve pi-pi interactions between the aromatic portion of the model glucuronides and the benzene ring of PBA (glucuronic acid itself is not extracted onto PBA under the conditions employed here, Tugnait *et al* unpublished observations). Clearly however, the structure of the analyte is important, and not just the possession of a number of aromatic rings, given the differences in extraction onto PBA exhibited by NG (poor) and BNDG (complete). Further studies are obviously needed to determine the precise structural requirements for efficient extraction of glucuronides onto PBA. It should be noted that the requirement for extraction at alkaline pH may militate against the use of PBA for ester glucuronides given their susceptability to base hydrolysis.

As noted in our previous studies (1-5) the C_{18} phase provides a more general means for the extraction of glucuronides but lacks the selectivity of PBA.

CONCLUSIONS

Both PBA and C_{18} bonded silica phases can be used to rapidly and efficiently concentrate, isolate and purify glucuronides from urine. The C_{18} material provides a general method for this class of metabolites, of low specificity, whilst the PBA phase appears to

Table 1. Degree of retention of the model glucuronides for different sorbent capacities. Number of symbols indicates magnitude of retention: + = < than 20%, ++ = slightly > than 90% and ------ = not investigated.

GLUCURONIDE	MOL. WT.	DEGREE OF RETENTION FOR DIFFERENT SORBENT CAPACITIES						
		100mg	200mg	300mg	400mg	500mg	600mg	1000mg
Phenolphthalein glucuronic acid	516.4	+	++	+++	Total retention	Total retention	----	----
6-Bromo-2-Naphthyl-β-D-glucuronide	399.2	----	----	----	----	+++	Total retention	----
6-Nitrophenyl-β-D-glucuronide	315.2	+	----	----	----	+	----	+
6-Naphthyl-β-D-glucuronide	319.3	----	----	----	----	++	----	++

provide high selectivity for a more limited range of compounds. In the case of extraction of glucuronides onto PBA the possession of the glucuronic acid moeity is insufficient in itself to ensure extraction and the structure of the aglycone is also clearly important.

REFERENCES

1. J.K. Nicholson and I.D. Wilson. High resolution proton magnetic resonance spectroscopy of biological fluids. *Prog. NMR Spectrosc.* 21, 449-501 (1989).
2. I.D. Wilson and J.K. Nicholson. Solid-phase extraction chromatography and nucelar magnetic resonance spectrometry for the identification and isolation of drug metabolites in urine. *Anal. Chem.* 59, 2830-2832 (1987).
3. I.D. Wilson and J.K. Nicholson. Solid-phase extraction chromatography and NMR spectroscopy (SPEC-NMR) for the rapid identifcation of drug metabolites in urine. *J. Pharm. Biomed. Anal.* 6, 151-165 (1987).
4. I.D. Wilson and I.K Ismail. A rapid method for the isolation and identification of drug metabolites from human urine using solid phase extraction and proton NMR spectroscopy. *J. Pharm. Biomed. Anal.* 4, 663-665 (1986).
5. M. Tugnait, F.Y.K. Ghauri, I.D. Wilson and J.K. Nicholson. NMR monitored solid-phase extraction of phenolphthalein glucuronide on phenylboronic acid and C_{18} bonded phases. *Pharm. Biomed. Anal.* 9, 895-900.
6. H.L. Weith, J.L. Wiebers and P.T. Gilham. Synthesis of cellulose derivatives containing the dihydroxy-boronyl group and a study of their capacity to form specific complexes with sugars and nucleic acid components. *Biochemistry* 9, 4396-4401 (1970).
7. J.L. Rosenberg and P.T. Gilham. Studies on the interactions of nucleotides, polynucleotides and nucleic acids with dihydroxyboronyl substituted celluloses. *Biochemistry* 11, 3623-2628 (1972).

A NOTE ON SOLID PHASE EXTRACTION OF THE HIGHLY PROTEIN BOUND

DOPAMINE UPTAKE INHIBITOR GBR 12909

Steen H. Ingwersen and Lars Dalgaard

Pharmacokinetics
CNS Division
Novo Nordisk A/S
DK-2760 Maaloev
Denmark

INTRODUCTION

Solid phase extraction (SPE) in some cases yields poor recovery of highly protein bound drugs, and this phenomenon has been explained by insufficient retention of the analyte on the sorbent [1]. The problem may be surmounted by changing the pH of the sample, by addition of protein denaturing agents such as urea or by precipitating the proteins from the sample [1,2].

GBR 12909 is a disubstituted piperazine-derivative (Fig. 1) with potential antidepressive effects. It is a basic, lipophilic compound and is more than 99% protein bound in human serum. In initial SPE experiments, in which the sample was prediluted in water or applied undiluted to the extraction columns, recoveries from human serum were below 50%. However, recoveries were significantly improved by prediluting the sample in 8 M urea.

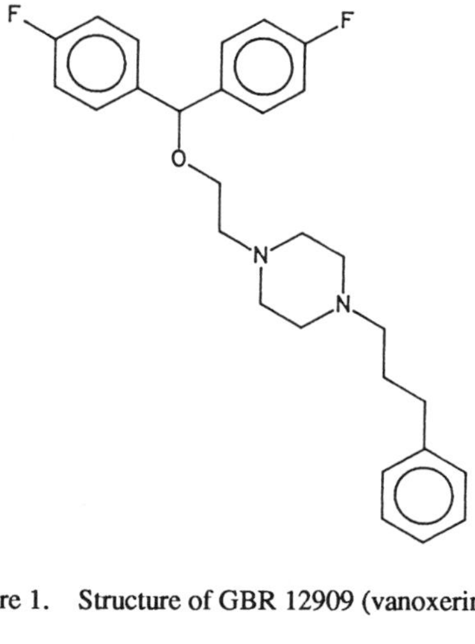

Figure 1.   Structure of GBR 12909 (vanoxerine).

Sample Preparation for Biomedical and Environmental Analysis,
Edited by D. Stevenson and I.D. Wilson, Plenum Press, New York, 1994

The purpose of the present study was to investigate the mechanism of action of urea as a sample diluent in SPE. Surprisingly, the effects of urea were shown not to be due to elimination of protein binding of the analyte.

EXPERIMENTAL

Chemicals And Reagents

GBR 12909 and internal standard [3] were synthesized by Novo Nordisk (Bagsvaerd, Denmark). Methanol of Chromoscan[R] grade was from Lab-Scan (Dublin, Ireland). Bond-Elut C_{18} SPE columns (2.8 ml, containing 500 mg of stationary phase, lot no. 73023 unless otherwise indicated) were from Analytichem International (Harbor City, CA, USA). Analytical grade urea, ammonia, sodium phosphate, sodium acetate and hydrochloric acid were from Merck (Darmstadt, F.R.G.). Tris HCl was from Sigma (St. Louis, MO, USA).

Solid-Phase Extraction

Two different sample eluents (methanol and 1% (v/v) ammonia in methanol) were investigated in combinations with the following sample diluents: water, urea (8 M), 0.1 M sodium acetate buffers (pH 4.0 and 5.0), 0.1 M sodium phosphate buffers (pH 6.0, 7.0 and 7.8) and 0.1 M tris buffer (pH 9.0). Otherwise, the SPE procedures were identical throughout.

All elutions of extraction columns were performed by low-speed centrifugation (128 x g, $18^{\circ}C$) except for the last washing with 60% (v/v) methanol, which was performed at 2000 x g with slow acceleration. The columns were conditioned with 2 ml of methanol followed by 2 ml of water. Then 0.5 ml of serum was pipetted directly onto the top of the columns followed by 0.1 ml of internal standard and 0.5 ml of sample diluent. The columns were subsequently washed with 2 ml of water and 2 ml of 60% (v/v) methanol in water. The columns were finally eluted by 2 ml of sample eluent.

Extracts were evaporated to dryness, redissolved in methanol-water (60:40) and assayed by HPLC as described [3]. Analyte recoveries were estimated relative to a standard curve obtained using unextracted calibrators prepared in the HPLC eluent.

RESULTS AND DISCUSSION

Fig. 2 shows that recoveries of GBR 12909 from human serum by different sample diluents was highly dependent on the choice of sample eluent. When eluted with pure methanol, the recovery increased considerably with 8 M urea as diluent in stead of water. However, this difference between diluents was absent when ammonia (1%) was included in the sample eluent (Fig. 2 left). The difference between pure methanol and ammonia in methanol as eluents was even more pronounced when buffers of different pH were used as sample diluents (Fig. 2 right). Thus, when eluting with methanol, the recovery decreased considerably with decreasing pH, whereas the pH effect was almost eliminated with ammonia-methanol as eluent.

These results clearly indicate, that problems with low recovery of this analyte was not due to lack of retention and thus the effect of urea could not be explained as an effect on protein binding. Rather, the effects of urea (and ammonia) were due to influences on secondary interactions between the analyte and free silanol groups present on the sorbent. Thus, secondary interactions between basic compounds and bonded silica surfaces are reduced in basic solutions as a result of reduced protonation of the analyte [4]. In basic solutions with pH above 8, GBR 12909 - having pK_a's of 4.0 and 8.0 - predominantly exists as the unprot-

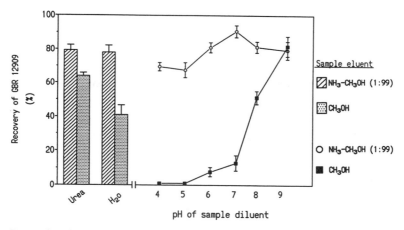

Figure 2.   Recoveries of GBR 12909 by solid phase extraction of human serum using different sample diluents and sample eluents. Results are mean ± SD (n=6).

onated component. This agrees well with the high recoveries obtained using tris buffer, pH 9.0 as sample diluent, regardless of the nature of the sample eluent (Fig. 2). Likewise, urea and ammonia may reduce the secondary interactions merely by reducing the protonation of the analyte. Alternatively, urea interacts directly with the silica surface and thereby eliminates secondary interactions between analyte and free silanol groups.

The involvement of secondary interactions was further supported by the lot-to-lot variations seen with SPE columns. Fig. 3 shows that the poor recovery with urea as sample diluent and pure methanol as eluent seen in the above experiments with one lot of columns (no. 73023) was not seen in three other lots tested. The extent of free silanol groups present is one of the parameters that may vary between lots of bonded silica, and the present results show that the ruggedness towards different SPE column lots should be evaluated when validating a new SPE procedure.

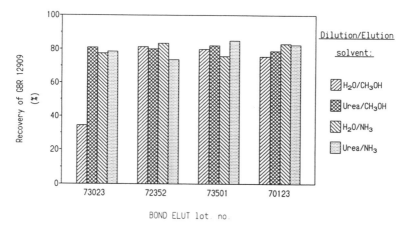

Figure 3.   Recoveries of GBR 12909 from human serum using four different lots of SPE columns with different combinations of sample diluents and sample eluents.

CONCLUSIONS

The poor recoveries of GBR 12909 obtained with some lots of SPE columns was improved by diluting the sample with 8 M urea or with 0.1 M tris buffer, pH 9.0 instead of water or by eluting the analyte with ammonia-methanol. The effect of urea was not due to reduced protein binding but more likely to reduced secondary interactions.

ACKNOWLEDGEMENT

The technical assistance of Y. Petersen is highly appreciated.

REFERENCES

1. K.C. Van Horne, ed., Sorbent Extraction Technology. Analytichem International, Harbor City (1985).
2. P.G.M. Zweipfenning, J.A. Lisman, A.Y.N. Van Haren, G.R. Dijkstra and J.J.M. Holthuis, Determination of Δ^9-tetrahydrocannabinol in plasma using solid-phase extraction and high-performance liquid chromatography with electrochemical detection, *J. Chromatogr.* 4560:83 (1988).
3. S.H. Ingwersen, Column liquid chromatographic assay of the dopamine reuptake inhibitor vanoxerine (GBR 12909) in human serum, *J. Chromatogr* 571, 305-311 (1991).
4. D.C. Leach, Reversed-phase HPLC of basic samples, *LC-GC International* 1, 22-30 (1988).

A NOTE ON THE ANALYSIS OF VALPROIC ACID IN HUMAN PLASMA USING AN

AUTOMATED SAMPLE PREPARATION DEVICE (ASTED)

A.D. Dale* and S.E Turner

Biokinetix Limited
Potters Bar
Hertfordshire, UK

INTRODUCTION

Valproic acid (VPA), the structure of which is shown as Figure 1, was first synthesised by Burton in 1862 [1], but its pharmacological importance remained undiscovered until 1963, when Meunier [2] showed, almost by accident, that VPA has anti-convulsant activity, with a broad spectrum of anti-epileptic activity. The mechanism of action of VPA is not fully understood, but it is thought to be related in some way to altered levels of gamma aminobutyric acid (GABA), which is an inhibitory neurotransmitter in the brain [3].

Several methods have been described for the analysis of VPA in plasma and other biological fluids. These include gas liquid chromatography (GC) [4,5] high pressure liquid chromatography (HPLC) [6,7] and several immunoassay techniques [8].

Valproic acid is volatile and special methods have been introduced to ensure that it is not lost during the extraction steps. VPA exhibits no significant UV absorption so it is common practice to derivatise the compound prior to analysis. This enables a higher wavelength to be used for detection, and also reduces the possibility of loss of VPA prior to analysis, but it does increase the overall analysis time.

The method described here overcomes these difficulties by using an ASTED (Automated sample, trace enrichment, dialysis) sample processor to isolated the VPA, at therapeutic levels, from human plasma. The assay is full automated and the extraction proceeds concurrently with the chromatographic analysis, thus reducing overall analysis time.

$$CH_3 - CH_2 - CH_2 - \underset{\underset{CH_2 - CH_2 - CH_3}{|}}{CH} - COOH$$

Figure 1. The structure of valproic acid.

---

* Present address: Resolution Chemicals, Stevenage, Herts. UK.

Sample Preparation for Biomedical and Environmental Analysis,
Edited by D. Stevenson and I.D. Wilson, Plenum Press, New York, 1994

PRINCIPLE OF THE ASTED

The ASTED performs the following operations automatically:

i)   The sample, with an internal standard if necessary, is transferred to one side of a semi-permeable membrane with a molecular weight cut off of about 15,000. On the other side of the membrane is a recipient solution, the composition of which is varied according to the application, and into which the drug and internal standard diffuse.

ii) After allowing this diffusion, the recipient solution is pumped through a short column packed with octadecyl bonded silica, which retains the compounds of interest.

iii) This short column is then placed in line with the analytical column, using conventional column switching techniques, and the various retained compounds are desorbed from the short column and transferred to the top of the analytical column as a tight band.

EXPERIMENTAL

Materials And Methods

Valproic acid (sodium salt) and diltiazem hydrochloride, which was used as the internal standard, were supplied by Generics (UK) Limited. Other reagents used were of analytical grade and supplied by FSA (Loughborough, UK). The water was deionised by an Elga reversed osmosis water purification system (Elga, High Wycombe, Buckinghamshire, UK). The ASTED was supplied by Anachem (Luton, Bedfordshire, UK) and was fitted with a 15K-Dalton semi-permeable membrane. The trace enrichment cartridge was packed with C18-bonded silica. The chromatograph was modular, consisting of a Spectra-Physics solvent pump (Spectra-Physics, Hemel Hempstead, Hertfordshire, UK), and a Spectra-Physics variable wavelength UV detector Model 8773 (Spectra-Physics). The output from the detector was fed to a Spectra-Physics Model 4200 recording integrator (Spectra-Physics). The column used was Nucleosil C18 (15 cm x 4.6 mm) (FSA, Loughborough) with a LiChrosorb C18 precolumn (FSA).

ASTED OPERATING CONDITIONS

The ASTED operating conditions were as follows. The trace enrichment cartridge contained 10 µm Hypersil ODS. The dialyser recipient solvent was potassium dihydrogen orthophosphate buffer (10 mm) at pH 5.0. The primary solvent was 0.86 (w/v) sodium chloride in water. The method file was as follows: sample volume, 500 µl; air gap volume, 5; pulse mode, 0; load volume, 600 µl; enrichment volume, 400 µl and trace enrichment cartridge elution time, 50 seconds.

CHROMATOGRAPHIC CONDITIONS

Chromatography was performed on a Nucleosil 5 µm ODS column (15 cm x 4.6 mm i.d.) with a mobile phase consisting of acetonitrile-potassium phosphate buffer (10 mm, pH 4) 45:55, at a flow rate of 1.5 ml/min. Detection was at 210 nm at 0.001 AUFS.

RESULTS AND DISCUSSION

A typical chromatogram, showing the results obtained for a sample containing 50 µg/ml, is shown in Figure 2. The results showed a good linear relationship between the ratio of the drug to the internal standard and the concentration of the valproic acid added. This

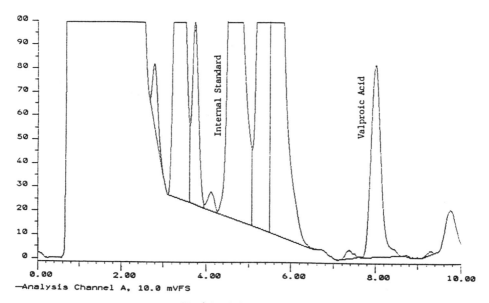

Figure 2. A typical chromatogram of valproic acid extracted from human plasma using the ASTED conditions described in the text.

relationship holds at least over the range of 5 μg to 200 μg/ml in human plasma.

This provides excellent data over the therapeutic ranges likely to be encountered with the normal valproic acid dosage [9].

CONCLUSION

A method has been described for the analysis of valproic acid in human plasma in the therapeutic range of 5 μg/ml to 200 μg/ml. The new method has no pre- or post-column derivatisation steps and is consequently more rapid than previously published methods. Experiments are continuing to improve the lower limit of detection to enable the method to be used for bio-availability studies in addition to therapeutic monitoring.

REFERENCES

1. B.S. Burton. (Unavailable). *Amer. Chem. J.*, 3, 385-395 (1882).
2. H. Meunier, G. Carraz, Y. Meunier, P. Eymard and M. Aimard. Pharmacodynamic properties of N-dipropylacetic acid. *Therapie.*, 18, 435-438 (1963).
3. G. Carraz, M. Darbon, S. Lebreton and H. Beriel. Pharmacodynamic properties of N-dipropylacetic acid. *Therapie.*, 19, 468-476 (1964).
4. F. Andreolin, C. Borra, A. DiCorcia and R. Samperi (1984). Direct determination of valproate in whole blood samples. *J. Chromatogr. (Biomed. Applic.)*, 310, 208-212 (1984).
5. I.M. Vonk and M. Beeksma. Rapid G.C. determination of valproic acid in serum. *Inter. Lab.*, 30-35 (1988).
6. L.J. Lovett, G.A. Nygard, G.R. Erdmann, C.Z. Burley and S.K. Wahba Khalil. H.P.L.C. determination of valporic acid in human serum using ultra-violet detection. *J. Liquid Chromatogr.*, 10, 687-699 (1987).
7. R.N. Gupta, P.M. Keane and M.L. Gupta. Valproic acid in plasma, as determined by liquid chromatography. *Clin. Chem.*, 25, 11-21 (1979).
8. J.T. Burke and J.P. Thenot. Determination of Antiepileptic drugs. *J. Chromatogr.*, 340, 199-241 (1985).
9. Martindale - The Extra Pharmacopoeia (28th Ed.) J.E.F. Reynolds (ed.), The Pharmaceutical Press, London, 1256-1259 (1982).

A NOTE ON PROBLEMS AND SOLUTIONS ENCOUNTERED DURING METHOD DEVELOPMENT FOR THE ANALYSIS OF NISOLDIPINE, A SECOND GENERATION CALCIUM ANTAGONIST

Sophia Monkman, Suzanne Cholerton and Jeffrey Idle

Pharmacogenetics Research Unit
Department of Pharmacological Sciences
University of Newcastle upon Tyne
The Medical School
Framlington Place
Newcastle, NE2 4HH, UK

INTRODUCTION

Heart disease is the primary cause of premature death in the western world. Calcium antagonists can be used to combat many cardiovascular diseases, including coronary vasospasm and congestive heart failure. The dihydropyridine calcium antagonists are potent coronary and peripheral vasodilators and are widely used to ameliorate essential hypertension and chronic stable angina.

Nifedipine (Structure 1) was the first dihydropyridine drug to find clinical use and is the parent of many second-generation compounds. The latter are quite diverse but retain the dihydropyridine ring structure which confers pharmacological activity. Nisoldipine (Structure 2) differs from nifedipine only in having an additional 2-methylpropyl side-chain moiety, but is the most potent dihydropyridine drug currently available [1]. The rapid and extensive hepatic clearance of nisoldipine results in very low plasma concentrations. A typical maximal concentration after a single 10 mg dose is only 1.8 ng/ml. In order to study the pharmacokinetics of nisoldipine under various conditions a method capable of measuring between 0.05 and 10.0 ng/ml in plasma is needed.

Dihydropyridine drugs are sensitive to light with a wavelength below 450 nm [2] breaking down to a pyridine derivative (half life of photodegradation 11.3 min [3]). In the case of both nifedipine and nisoldipine this breakdown product is the same as one of the plasma metabolites. Thus the assay of nisoldipine has to be carried out under yellow light, using amber glass for all glassware and autosampler vials. Furthermore, neither fluorimetric nor spectrophotometric detection methods can be employed without breakdown of the compound. Therefore capillary gas chromatography (GC) with electron-capture detection (ECD) was chosen for this assay since detection of low concentrations of nisoldipine is possible utilising the separation power of capillary GC and the high response of the nitro-group to ECD. Injection technique is crucial because of thermal degradation. However, this

Sample Preparation for Biomedical and Environmental Analysis,
Edited by D. Stevenson and I.D. Wilson, Plenum Press, New York, 1994

NIFEDIPINE

NISOLDIPINE

Structure 1. Nifedipine   Structure 2.   Nisoldipine

can be minimised by using cool on-column injection into a wide-bore capillary pre-column with a very thin film coating.

The established assay for nifedipine has a limit of detection of only 5 ng/ml [4] and as this was not sufficient for our nisoldipine studies investigations to purify and concentrate the sample were made.

EXPERIMENTAL

Materials

Nisoldipine (2-methylpropyl-methyl-1,4-dihydro-2,6-dimethyl-4-(2-nitrophenyl)-3,5-dicarboxylate) was a gift from Bayer, UK. [^{14}C]-nisoldipine was a gift from Bayer, Germany. Inorganic reagents were Analar grade and organic solvents were Distol grade or re-distilled Analar grade. Solid-phase extraction cartridges were either Bond Elut (Jones Chromatography Ltd, Hengoed, UK) or Baker (Linton Products, Diss, UK). Amber glass vials (4 ml, PTFE-faced solid caps) were from Zinsser Analytic (UK) Ltd (Maidenhead, UK) and tapered autosampler vials (500 µl, amber glass) were from Chromacol Ltd (London, UK). Plasma was obtained from the Regional Blood Transfusion Service, (Tyneside, UK). Scintillation fluid was from LKB (FSA, Loughborough, UK).

Methods

Solid-phase extraction (SPE) was carried out using a Jones Elution Manifold (Jones Chromatography Ltd) connected to a Capex 2DC oil-free pump (Charles Austen Pumps Ltd, Weybridge, UK) via a cold solvent trap (conical side-arm flask in ice). The pump effluent was piped to a filter cabinet (Bigneat Ltd, Havant, UK) where all solvents were handled. Liquid-liquid extractions were achieved by either tumbling on a rotary wheel (Harvard/Lte, BDH Laboratory Supplies, Merck Ltd, Poole, UK) or whirlimixing on a customised multiple-sample vortex mixer (Vibrax-VXR, IKA Labortechnik, Janke & Kunkel, Germany). Scintillation counting was achieved using a Packard Tri-Carb 4530 (Canberra Packard, Pangbourne, UK).

Chromatographic analysis was carried out using a Hewlett-Packard 7673A robotic autosampler which injected 0.2 µl of toluene extract into the on-column injection port of a Hewlett Packard 5890A gas chromatograph equipped with a ^{63}Ni ECD (at 300°C) (15 mCi). Injection was achieved directly into a wide-bore capillary pre-column (7 m x 0.53 mm i.d. x 0.1 µm DB1, J & W Scientific, California, USA) glass butt-connected to an analytical column (40 m x 0.25mm i.d. x 0.25 μm DB1). The injection port was unheated and chromatography was carried out isothermally at 255°C. Carrier gas was hydrogen at 35 psi back pressure (ca. 90 cm/sec) with a nitrogen make-up flow of 40 ml/min. A 3396A computing integrator was used to control the autosampler and to process the chromatographic signals. Under the described conditions the retention time for nisoldipine was 7.99 min.

Reversed-Phase SPE

Using the Bond Elut sorbents C-18, C-8, C-2, CH, PH the retention of nisoldipine from aqueous solution was assessed. Organic solvents (acetone, acetonitrile, diethyl ether, methanol, methyl tert-butyl ether and propan-2-ol) were then tested for elution efficiency and for cleanliness of extract.

The sorbents and solvents chosen from the above were tested for retention and elution respectively, of nisoldipine in plasma. Follow-up investigations were made into the effect upon retention of adding various buffers to the plasma before extraction, and into the effect of aqueous/organic solvent mixtures as rinse solutions to remove plasma inteferents.

Multistage Liquid-Liquid Extraction

Using [14C]-nisoldipine, a multistage liquid-liquid extraction method was attempted [5]. Britton-Robinson-type buffers [6] with pH values from 0.1 to 12.0 were prepared and were spiked with [14C]-nisoldipine (1.0 mg/l). Solvents were chosen which had suitable physiochemical characteristics [7] (e.g. immiscibility with water and appropriate polarity) and liquid-liquid extractions (1 hr on rotary wheel) were carried out with aqueous buffer:organic solvent (1:2). An aliquot (500 μl) was taken from each phase for scintillation counting.

Liquid-Liquid Extraction Coupled To Normal-Phase SPE

[14C]-nisoldipine in toluene or octanol was passed through a range of SPE cartridges (see Table 2 for details) and the waste collected to assess retention. The SPE cartridges were primed with methanol (two column-volumes), propan-2-ol (two column-volumes) and octanol or toluene (two column-volumes).

Liquid-liquid extraction was carried out, as previously described, using unlabelled nisoldipine in plasma. After centrifugation the toluene extract was passed through the SPE cartridge chosen. Toluene (one column volume) was passed through to remove plasma inteferents and the nisoldipine eluted with propan-2-ol. This extract was dried by centrifugal evaporation, reconstituted with toluene and injected onto the GC to assess recovery and cleanliness of extract. The liquid-liquid and solid-phase extractions were subsequently optimised.

RESULTS AND DISCUSSION

Reversed-Phase SPE

The first problem encountered when using the reversed-phase SPE methodology for nisoldipine in plasma was the low recovery due to extensive protein binding (>99%) of the drug [8]. The second difficulty was passing viscous plasma samples through the SPE cartridges. These problems were solved by diluting the plasma with buffer, thereby reducing protein binding and increasing retention, and allowing easier passage of the sample through the cartridge. However, addition of buffer did not remove the small clots of protein within

Table 1. Recovery of ^{14}C-nisoldipine after
liquid-liquid extractions using a
variety of organic solvents.

Solvent	Extraction from buffer (%)	Extraction from plasma (%)
Butanol	99	78
Cyclohexane	84	34
Heptane	83	18
Hexane	86	19
Hexanol	94	76
Pentane	100	44
Octanol	96	90
Toluene	98	94
Trimethylpentane	100	44

the plasma sample which blocked the cartridge and lengthened elution times greatly. Centrifugation of the sample prior to SPE was a doubtful solution as the extensive protein binding of nisoldipine meant that loss of protein clot could have meant loss of drug. Rinses using various solvent/water mixtures were devised to remove the co-extracted plasma interferences but the extract was still not clean enough and reversed-phase SPE was abandoned.

Multistage Liquid-Liquid Extraction

The extraction efficiency obtained for a ranged of solvents tested for extraction of [^{14}C]-nisoldipine from buffer and from plasma is shown in Table 1. The results shown for buffer represent the minimum recovery obtained over the entire pH range. There was no influence of pH upon extraction and a pKa value for nisoldipine could not be determined. Back-extraction into aqueous buffer and hence multistage liquid-liquid extraction was therefore impossible. Furthermore, the recovery from plasma was generally far less than from buffer. Only octanol and toluene were able to extract over 90% nisoldipine from plasma.

Liquid-Liquid Extraction Coupled To Normal-Phase SPE

Data for toluene only is presented since octanol proved unusable as it was almost impossible to pass it through the SPE cartridges due to its high viscosity. The retention efficiencies obtained with the SPE cartridges tested are shown in Table 2. Many authors [9-11] have commented upon differences in selectivity of the same phase from different manufacturers and it can be seen that the Bond Elut amino (NH$_2$) phase retains less than half the [^{14}C]-nisoldipine whilst the NH$_2$ phase from Baker retains it all. Normal phase retention of [^{14}C]-nisoldipine was greatest with the Bond Elut silica cartridges and the Baker NH$_2$ cartridges. The latter were chosen for further use as they gave a cleaner extract.

Sample Concentration Problems

The standard autosampler vial for the HP7673A has a volume of 2 ml. The minimum solvent volume that can be injected from within this vial is 400 µl. However, this sample

Table 2. Normal phase retention of [^{14}C]-ni-
soldipine by various solid-phase extraction cartridges.

Manufacturer	Cartridge	Retention
Bond Elut	CN	21
	NH2	44
	2OH	32
	Si	100
	C2	48
	C18	13
	CH	14
	PH	19
	Certify	49
J T Baker	NH2	106

preparation protocol was designed to concentrate from 1 ml plasma sample to a 100 µl toluene extract, thus rendering 2 ml autosampler vials unusable. However, 1 ml propan-2-ol was used for elution from the SPE cartridge, therefore a vial was needed that could be eluted into and injected from onto the GC. Although a tapered vial seemed to be the solution the only amber glass vial available had a volume of 500 µl. It was therefore necessary to attempt to reduce the elution volume from 1 ml to 500 µl.

Reduction of elution volume was achieved by addition of acetic acid (final concentration of 0.17 M) to the propan-2-ol pre-wash and eluant. The elution profiles achieved are shown in Figure 1 and demonstrate the effect of increased acid passed through the SPE cartridge. A clear sequence of increasingly focused elution and increased recovery can be seen. When the concentration of the acid was doubled (0.34 M) in both pre-wash and eluant a recovery of 90% was achieved within the desired elution volume of 500 µl.

FINAL ASSAY CONDITIONS

To plasma standard or unknown Tris buffer (0.2 M, pH 9.0, 200 µl) and toluene (2.5 ml) were added. After whirlimixing (1800 rpm; 20 min) the sample was centrifuged (2500 rpm; 10 min). Baker SPE cartridges (NH$_2$, 200 mg/3 ml) were prepared with two column volumes each of methanol, propan-2-ol (with 0.34 M acetic acid), and toluene, drawn through with little or no vacuum. After centrifugation of the sample the toluene supernatant was passed through the cartridge under gravity. The vacuum was raised to 5 mm Hg and toluene (one column volume) passed through quickly to remove loosely bound inteferent compounds. Vacuum was maintained (3 min) to dry the cartridge and then reduced (2 mm Hg) for elution of the sample. Using acidic propan-2-ol (500 µl) the sample was eluted into tapered amber glass vials (500 µl), dried by centrifugal evaporation (45°C, ca. 25 min), reconstituted in toluene (100 µl) and an aliquot (0.2 µl) injected onto the GC.

The chromatograms shown in Figure 2 are of the concentrated extract obtained and they indicate the difficulty in measuring low-level analytes in plasma. The lowest concentration to be accurately measured by this assay is 0.1 ng/ml and as can be seen from Figure 2 the peak for nisoldipine at this concentration (chromatogram B) is easily detected and clearly separated from plasma inteferents. The Baker SPE cartridges were chosen also because they were able to produce a clear region where nisoldipine elutes, amongst the myriad of plasma components.

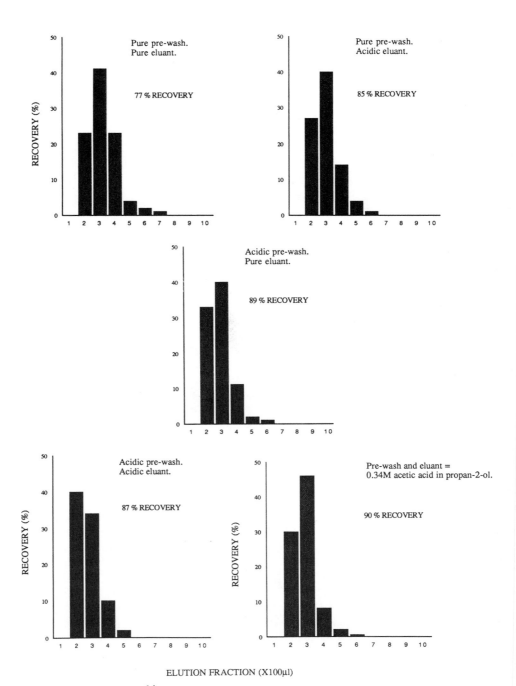

Figure 1. Elution profiles of [^{14}C]-nisolipine from Baker NH$_2$ SPE cartridges under conditions of increasing acidity.

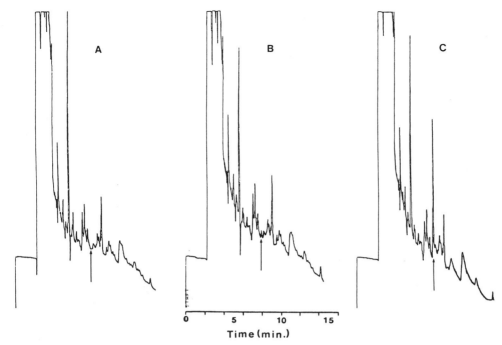

Time (min.)

Figure 2. GC-ECD chromatograms of: (A) blank plasma extract; (B) extract of plasma containing nisoldipine at a concentration of 0.1 ng/ml, and (C) extract of plasma containing nisoldipine at a concentration of 5.0 ng/ml. The arrow points to the region of the peak of interest.

CONCLUSION

This assay was developed to assay nisoldipine in plasma samples from clinical trials. It has fulfilled the assay requirements of detection from 10.0 to 0.05 ng/ml and will shortly be validated.

ACKNOWLEDGEMENTS

Thanks go the Bayer UK Ltd for financial support and to Robin Whelpton for valuable advice.

REFERENCES

1. R.A. Janis, A.V. Shrikhande, R. Greguski, M. Pan and A. Scriabine. Review of nisoldipine binding studies, in: "Nisoldipine", P.G. Hugenholtz and J. Meyer (eds.), Springer-Verlag, Berlin, pp27-35 (1987).
2. I.A. Majeed, W.J. Murray, D.W. Newton, S. Othman and W. Al-Turk. Spectrophotometric study of the photodecomposition kinetics of nifedipine. *J. Pharm. Pharmcol.*, 39, 1044-1046 (1987).
3. J. van Harten, M.T.M. Lodewijks, J.W. Guyt-Scholten, P. van Brummelen and D.D. Breimer. Gas chromatographic determination of nisoldipine and one of its metabolites in plasma. *J. Chromatogr.*, 423, 327-333 (1987).
4. B.J. Schmid, H.E. Perry and J.R. Idle. Determination of nifedipine and its three principal metabolites in plasma and urine by automated electron-capture capillary gas chromatography. *J. Chromatogr.*, 45, 107-109 (1988).
5. B.B. Brodie, S. Udenfriend and J.E. Baer. The estimation of basic organic compounds in biological material: I. General principles. *J. Biol. Chem.*, 168, 299-309 (1947).
6. W.C. Johnson and A.J. Lindsey. An improved universal buffer. *Analyst*, 64, 490 (1939).
7. C.R. Jones. Solvent extraction, in: "Assay of drugs and other trace compounds in biological fluids", E. Reid (ed.), North-Holland, Oxford, pp107-113 (1976).

8. G. Ahr, W. Wingender and J. Kuhlmann. Pharmacokinetics of nisoldipine, **in**: "Nisoldipine", P.G. Hugenholtz and J. Meyer (eds.), Springer-Verlag, Berlin, pp59-66 (1987).

9. R.P.W. Scott and P. Kucera. Examination of five commercially available liquid chromatographic reversed phases (including the nature of the solute-solvent-stationary phase interactions associated with them), *J. Chromatogr.*, 142, 213-232 (1977).

10. R.J Ruane and I.D. Wilson. The use of C18 bonded silica in the solid phase extraction of basic drugs - possible role for ionic interactions with residual silanols. *J. Pharm. Biomed. Anal.*, 5, 723-727 (1987).

11. H. Salari and S. Steffenrud. Comparative study of solid phase extraction techniques for isolation of leukotrienes from plasma. *J. Chromatogr.*, 378, 35-44 (1986).

A NOTE ON THE USE OF NOVEL BONDED PHASE ADSORBENTS FOR THE EXTRACTION OF DRUGS OF ABUSE FROM BIOLOGICAL MATRICES

E. Heebner[1], M. Telepchak[1] and D. Walworth[2]

[1] Worldwide Monitoring Corporation
417A Caredean Drive
Horsham
PA 19044
USA

[2] Technicol Ltd
Brook Street
Higher Hillgate
Stockport
SK1 3HS. UK

INTRODUCTION

State-of-the-art extraction technology for drug of abuse analysis has its origins in liquid-liquid extractions. The literature is filled with liquid-liquid methods for extracting a wide variety of drugs with an even wider variety of methods. Some methods are effective, some are not [1-4]. Major difficulties with liquid-liquid extractions are that highly polar, water soluble molecules are difficult to extract with high efficiency from biological matrices. Methods are time consuming and large volumes of solvents are required, the resulting extracts are not necessarily clean and recoveries may be variable. The first steps towards alternative methods which would eventually result in the ultimate development of solid-phase extraction (SP) involved the use of columns based on diatomaceous earth.

The column itself was little more than a sophisticated liquid-liquid column in which the buffered aqueous fraction of a biological sample was adsorbed onto the diatomaceous earth particle and formed a stationary liquid phase [5,6]. A specified volume of organic solvent such as methylene chloride was then passed down the tube. Too little solvent resulted in loss of recovery. Too much solvent resulted in the elution of interfering peaks.

The next step on the evolutionary ladder came in the early 1980's when researchers began trying to use conventional hydrophobic bonded-phase extraction columns such as C18 and XAD_2 (cross-linked polystyrene-divinyl benzene) resins for the extraction of drugs of abuse. These were commonly used in therapeutic drug monitoring at that time [7-10]. The strategy of use with these materials to effect an extraction for drugs of abuse was to raise the pH to between 8 and 9 to deprotonate amines while attempting to extract the neutral compounds and ionised acids on the long hydrocarbon chain. This methodology was plagued by two problems. First, the same characteristics which make hydrophobic columns good for doing chromatography on a wide variety of compounds makes them bad for extraction work. Not only do they extract compounds of interest, but they also extract many interfering compounds at the same time. The second problem is that the degree of ionisation of a

Sample Preparation for Biomedical and Environmental Analysis,
Edited by D. Stevenson and I.D. Wilson, Plenum Press, New York, 1994

155

compound, which is pH dependent, affects its recovery. Assay recovery of different drugs can vary from less than 50% to over 99% using the same method. The use of ion exchange columns quickly followed [11,12]. Cleanliness and recovery improved greatly for charged drugs, but ion exchange columns cannot be effectively used for the analysis of neutral drugs.

By the mid 1980's, various external factors began to shape the demands placed on the next generation of techniques used for extracting drugs of abuse. These factors include increased pressure for drugs of abuse testing, the use of more sensitive, legally defensible analytical techniques for both qualitative and quantitative analysis, the development of more potent drugs used at lower levels, the broad chemical ranges of drugs being abused, and the need for removing more background interference from the sample.

Analysts had defined their demands and challenged the manufacturers to produce bonded-phases which:

1) were reproducible from lot to lot;
2) were easy to use;
3) processed many samples quickly;
4) reduced chromatographic background noise;
5) gave high, quantitative recovery of compounds;
6) worked over a wide chemical range of compounds;
7) were cost effective.

This lead to the development of copolymeric bonding techniques to yield what are now referred to as "designer" phases [13-21]. Extraction strategy changed as these new materials were developed. Instead of using harsh conditions around pH9, a gentler approach was used. The copolymer which was developed for this purpose is a hydrophobic cation exchange resin called "Clean Screen".

THE NATURE OF CLEAN SCREEN

The ideal use of solid-phase extraction is dependent upon two unique events: the retention of all compounds of interest on the extraction column, and the selective elution of these compounds in a purified state. Consequently, using conventional extraction columns there was often a trade-off between recovery and selectivity. Clean Screen extraction columns were designed to be consistent, rugged, easy to use copolymers allowing the deliberate use of mixed mode interactions (rather than the avoidance of them). Such columns exhibit hydrophobic properties consistent with approximately a C8 chain length and the cation exchange strength of a benzene sulfonic acid functional group.

Table 1.   Recoveries of a range of analytes from "Clean Screen" copolymer cartridges.

Analyte	% Recovery (Mean±SD)	Analyte	% Recovery (Mean±SD)
Amobarbital	98.1±2.9	Imipramine	98.0±0.4
Amphetamine	98.0±2.9	Meperidine	99.2±1.8
Butabarbital	98.8±3.0	Methadone	93.5±1.7
Cocaine	98.5±1.9	Methamphetamine	100.5±3.1
Codeine	97.4±3.6	Pentobarbital	96.5±2.4
Glutethimide	98.3±2.7	Phenylbutazone	97.0±0.4
		Phenytoin	96.0±2.6

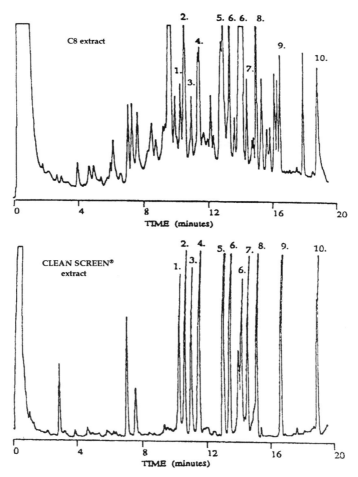

Figure 1. Chromatograms showing differences between samples prepared on a C8 hydrophobic bonded phase column and a Clean Screen copolymeric column. Urine samples (5 ml). Peak identification: 1. Butabarbital, 2. Amobarbital, 3. Pentobarbital, 4. Secobarbital, 5. Clutethimide, 6. Caffeine and metabolites, 7. Phenobarbital, 8. Alphenal (ISTD), 9. Methaqualone, 10. Phenytoin.

Several years of studying the variables in the copolymerisation process has enabled the production of a phase which does not suffer from the common problem of lot to lot reproducibility seen in many current solid-phase extraction products. The effectiveness of using the proper copolymer to create a surface which yields high recoveries of drugs of a broad chemical nature is shown in Table 1.

EXTRACTION METHODOLOGY

When a sample is loaded onto the column at pH 6, the carboxylic acid functionalities present in the sample are ionised. This creates a repulsion between the column and many sample borne interferences, thereby reducing the likelihood of their adsorbing onto the column. At this pH, barbiturates and methaqualone are not ionised and are hydrophobically adsorbed onto the column. At the same time, drugs with amine functions such as cocaine, and opiate drugs adsorb onto the column via both hydrophobic and ionic attraction.

The column can then be washed with water or weak aqueous buffers at or below pH

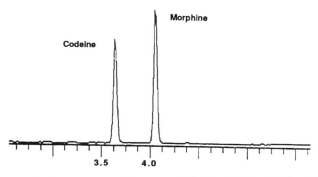

Figure 2.   GC/MS selected ion chromatogram of ion masses 371, 234 and 343 of TMS-codeine and 287, 324 and 429 of TMS-morphine extracted from urine spiked at 360 ng/ml.

6 without risking loss of the analytes. When washed with 1.0 M acetic acid, pH 2.3, metha-qualone will become cationic and therefore ionically bound to the column. After drying the column, it is possible to elute the hydrophobically bound analytes using solvents of minimal polarity such as methylene chloride or a hexane-ethyl acetate mixture. Cationic analytes which remain bound to the column can be eluted after another drying step. Many com-pounds of intermediate polarity, potential interferences, will also remain on the column. The majority of these potential interferences can be removed by using a methanol wash. The drying steps were necessary to remove water which would have prevented the water-immis-cible elution solvents from optimally interacting with the analytes.

The cationic analytes are finally eluted using a methylene chloride-isopropanol-ammonium hydroxide mixture, which simultaneously disrupts hydrophobic and ionic inter-actions. The advantage of this scheme becomes readily apparent in the chromatograms shown in Figure 1 where conventional C8 column and the "Clean Screen" are compared. The sample extracted on the Clean Screen extraction column had much less background chromotographic noise than that extracted using an endcapped C8 column.

Analyses of some compounds require special procedures for optimal recovery. The analysis of compounds with a carboxylic acid functional group often necessitates lowering the sample pH to below 6 at the time of sample loading to reduce anionic repulsive forces and minimise analyte loss at this step. Sometimes analyte polarities dictate the use of special elution solvents for the acid-neutral drug fraction. Use of the wrong solvent may significant-ly reduce recovery. Another notable special case is the analyses of glucuronide conjugates of drugs. Most frequently those drugs are hydrolysed by enzymatic, strong acid or base digestion prior to extraction to allow analysis by gas chromatography. Many glucuronidated

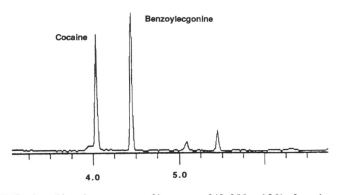

Figure 3.   GC/MS selected ion chromatogram of ion masses 240, 256 and 361 of cocaine and TMS-benzoylecgonine extracted from urine spiked at 180 ng/ml.

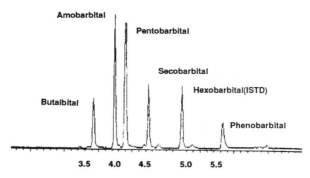

Figure 4.   GC/MS selected ion chromatogram of ion masses 153, 168 and 195 of barbiturates extracted from urine spiked at 200 ng/ml.

drugs with amine functionalities can be extracted without hydrolysis if some solvent changes are made to accommodate the higher polarity of these drugs.

The sample matrix may dictate special treatment of samples. With minor pre-extraction treatment, however, whole blood and tissue samples may be extracted on copolymeric extraction columns. Whole blood is usually diluted with water to lyse the red blood cells. After pH adjustment and centrifugation, the sample is ready for extraction. Tissue samples are prepared for extraction by homogenisation and centrifugation. Once the supernatant's pH is adjusted, it is ready for extraction. The majority of analyses for drugs of abuse are performed on urine specimens. These sometimes require centrifugation to remove precipitated salts, and hydrolyses are often used in samples suspected of containing cannabinoids, codeine or morphine to liberate the aglycone from the glucuronide. The usual preparation consists only of a simple pH adjustment and addition of an internal standard.

Applications Involving Gas Chromatography - Mass Spectrometry (GC-MS) With Selected Ion Monitoring Of Urine Extracts

Chromatograms of drugs of abuse which were extracted from commercially spiked urine and analysed by GC-MS are shown in Figures 2-5. These compounds, except for cocaine, were derivatised prior to injection. Each chromatogram shows data collected by monitoring three selected ions during data aquisition. One of these ions is used for quantification. The other ions are used to confirm the analyte's identity. The concentration spiked into each urine sample is 1.2 times the minimum level determined to be legally significant when tested by National Institute of Drug Abuse (NIDA) certified laboratories. Codeine and morphine were extracted from urine spiked at 360 ng/ml of urine (Figure 2). Cocaine and its metabolite benzoylecgonine were seeded at 180 ng/ml (Figure 3). Assay recoveries of these compounds exceeded 90% when extracted on the copolymeric cation exchange columns. A series of barbiturates were extracted from urine (Figure 4) which had been spiked at a level of 200 ng/ml. Their recoveries also exceeded 90%. The 11-nor-Δ9-carboxy-THC(THCA) chromatogram, shown in Figure 5 was from extracted urine which had been spiked at 18 ng/ml, also 1.2 times the NIDA "cut-off". This compound had also been derivatised prior to injection. When extracted using the copolymeric anion exchange column, the recovery exceeded 90%. THCA can also be extracted using a copolymeric cation exchange resin, but with lower recoveries.

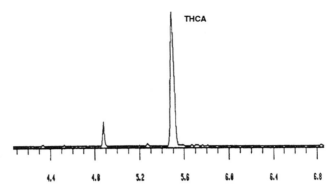

Figure 5.   GC/MS selected ion chromatogram of ion masses 371, 473 and 488 of TMS carboxy-THC ex
tracted from urine spiked at 18 ng/ml.

CONCLUSIONS

This new, mixed mode extraction technique enables the extraction and analysis a wide range of compounds, at low concentrations with minimum interferences from a wide range of biological matrices.

REFERENCES

1.   A.C. Moffat, J.V. Jackson, M.S. Moss and B. Widdop (eds.). Clarke's Isolation and Identification of Drugs, *The Pharmaceutical Press*, London, pp6-34 (1986).
2.   M.M. Baden, N.N. Valanju, S.K. Verma and S.N. Valanju. Confirmed identification of biotransformed drugs of abuse in urine. *Amer. J. Clin. Path.* 57, 43-51 (1972).
3.   L.J. Dusci and L.P. Hackett. The detection of some basic drugs and their major metabolites using gas-liquid chromatography. *Clin. Toxicol.* 14, 587-593 (1979).
4.   B.C. Thompson and Y.H. Caplan. *J. Anal. Toxicol.* 1, 66-69 (1977).
5.   J.D. Power, B. McKenna and M.B. Lambert. The use of extube extraction columns in horse drug screening. *Proceedings of the 7th International Conference of Racing Analysts and Veterinarians, Kentucky*, pp7-17 (1988).
6.   J.T. Stewart, T.S. Reeves and I.L. Honiberg. Comparison of solid-phase extraction techniques for assay of drugs in aqueous and human plasma samples. *Anal. Letts.* 17(B16), 1811-1826 (1984).
7.   N. Weissman, M.L. Lowe, J.M. Beattie and J.A. Demetriou. Screening method for detection of drugs of abuse in human urine. *Clin. Chem.* 17(9), 875-881 (1971).
8.   M. Stajic, Y.H. Caplan and R.C. Baker. Detection of drugs using XAD-2 resin. I. Choice of resin, chromatographic conditions and recovery studies. *J. Forensic. Sci.* 24(4), 722-731 (1979).
9.   N. Elahi. Encapsulated XAD-2 extraction technique for a rapid screening of drugs of abuse in urine. *J. Anal. Toxicol.* 4, 26-30 (1980).
10.   N. Dunnett, P. Chalmers, P.J. Bassett and M. Walden. A solid phase extraction method for general drug screening. *Proceedings of the International Conference of Racing Analysts and Veterinarians, Kentucky*, pp19-22 (1988).
11.   B.K. Logan, D.T. Stafford, I.R. Tebbett and C.M. Moore. Rapid screening for 100 basic drugs and metabolites in urine using cation exchange solid-phase extraction and high-performance liquid chromatography with diode array detection. *J. Anal. Toxicol.* 14, pp154-159 (1990).
12.   C. Moore. *JFSS* 30, 123-129 (1990).
13.   B. Thompson, J. Kuzmack, D. Law and J. Winslow. *LC-GC* 7, pp846-850.
14.   M.J. Kogan, D.J. Pierson, M.M. Durkin and N.J. Willson. Thin-layer chromatography of benzoylecgo-nine: a rapid qualitative method for confirming the EMIT Urine Cocaine Metabolite Assays. *J. Chromatogr.* 490, 236-242 (1989).
15.   Trinh Vu-Du and André Vernay. Simultaneous detection and quantitation of 06-monoacetylmorphine, morphine and codeine in urine by gas chromatography with nitrogen-specific and/or flame ionisation detection. *Biomed. Chromatogr.* 4(2), 65-69 (1990).

16. Trinh Vu-Du and André Vernay. Evaluation of sample treatment procedures for the routine identification and determination at nanogram level of 06-monoacetylmorphine in urine by capillary GC and dual NPD (nitrogen phosphorus detection) - FID. *Biomed. Chromatogr.* 13(3), 162-166 (1990).

17. R.L. Thies, D.W. Cowens, P.R. Cullis, M.B. Bally and Lawrence D. Mayer. Method for rapid separation of liposome-associated doxorubicin from free doxorubicin in plasma. *Anal. Biochem.* 188, 65-71 (1990).

18. L. Baskin, J. Charlson, P. Chezick and G. Lawson. Determination of serum nicotine concentration using solid-phase extraction and GC-MS with selected ion monitoring. *Amer. J. Clin. Path.* (1991).

19. J. Charlson and G. Lawson. Automated sample preparation and simultaneous analysis of four cardioactive drugs in serum. *To be published at A.A.C.C. Meeting in July 1991.*

20. L.C. Matassa, D. Woodward, R.K. Leavitt, P. Firby and P. Beaumier. Detection of glycopyrrolate in equine urine by LC/MS/MS and GC/MS. *Submitted for publication in J. Chromatogr. Biomed. Appl.*

21. A. Singh, M. Ashraf, K. Granby, U. Mishra and M. Mudhusudana Rao. Screening and confirmation of drugs in horse urine by using a simple column extraction procedure. *J. Chromatogr.*, 473, 215-226 (1989).

A NOTE ON MIXED MODE SOLID PHASE EXTRACTION OF BASIC DRUGS AND THEIR METABOLITES FROM HORSE URINE.

S.A. Westwood and M.C. Dumasia

Horseracing Forensic Laboratory Ltd
P.O. Box 15
Snailwell Road
Newmarket
Suffolk, CB8 7DT. UK

INTRODUCTION

Solid phase extraction (SPE) has become the method of choice for rapid, efficient sample clean up for the analysis of body fluids for drugs and metabolites. Various phases can be chemically bonded to silica support material producing a range of stationary phases similar to those used in high performance liquid chromatography (HPLC) including polar and non-polar phases, strong and weak anion or cation exchangers, etc.

Several manufacturers have produced solid phase extraction cartridges utilising dual retention characteristics. The combination of ion exchange with hydrophobic retention mechanisms is popular. The work described here was carried out using the Varian Certify or DAU Clean Screen SPE cartridges. These phases combine both strong cation exchange and hydrophobic retention characteristics. Previous work [1] has shown the utility of SPE cartridges of this type for analysis of β-agonists, β-antagonists and their metabolites in horse urine by capillary gas chromatography-mass spectrometry (GC-MS).

TYPICAL ANALYTICAL PROCEDURES

For basic drugs and metabolites a typical procedure employing these cartridges is as follows. Unconjugated drugs and metabolites are extracted directly, following adjustment of sample pH to 6.0. Conjugated drug metabolites are first hydrolysed. The sample is adjusted to pH 5.2 and *Helix pomatia* mixed enzymes added. The sample is then incubated overnight at 37°C. Basic metabolites are recovered by SPE using Certify or DAU Clean Screen cartridges as shown in Scheme 1.

Basic drugs and metabolites retained by the cartridge are eluted using ammoniacal ethyl acetate. The ammoniacal ethyl acetate used to elute the drugs and metabolites from the cartridge must be freshly prepared and vigorously shaken immediately prior to use, otherwise de-mixing can occur resulting in failed analyses.

Sample Preparation for Biomedical and Environmental Analysis,
Edited by D. Stevenson and I.D. Wilson, Plenum Press, New York, 1994

Scheme 1

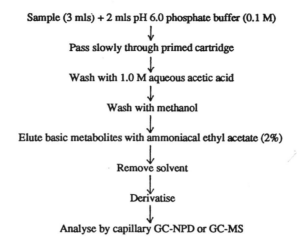

Sample (3 mls) + 2 mls pH 6.0 phosphate buffer (0.1 M)
↓
Pass slowly through primed cartridge
↓
Wash with 1.0 M aqueous acetic acid
↓
Wash with methanol
↓
Elute basic metabolites with ammoniacal ethyl acetate (2%)
↓
Remove solvent
↓
Derivatise
↓
Analyse by capillary GC-NPD or GC-MS

Instruments such as the Gilson ASPEC can be used to automate the procedure, however, the inherent instability of the eluting reagent causes problems. We have investigated the use of triethylamine as an alternative base to ammonia. Triethylamine (TEA) is readily miscible with many organic solvents, including ethyl acetate, which leads to a stable mixture and flexibility in the choice of solvent. This latter point can be important when an analyte is appreciably volatile, e.g. amphetamine, which would benefit from the use of a more volatile solvent such as diethylether. We have found that the use of TEA in the range 5 to 10% in ethyl acetate gives good results for a wide range of basic drugs and metabolites including acepromazine, atenolol, clenbuterol, isoxsuprine, levorphanol, metoprolol, promazine, propranolol, salbutamol, sotalol, terbutaline and timolol. An example of the GC-MS of a sample containing propranolol and isoxsuprine following extraction from plasma is shown in Figure 1.

APPLICATION TO ACIDIC METABOLITES OF BASIC DRUGS

Certain drugs, metoprolol and timolol amongst them, are extensively metabolised to carboxylic acid metabolites. The narcotic analgesic fentanyl (Sublimaze) is also reported to undergo phase I metabolism to a β-keto acid in the equine [2].

Metabolites of this type, having basic and acidic functionality, behave as zwitterions. These are particularly difficult to handle by conventional extraction methods, but can be analysed by SPE. At pH 5.5 the ionisation of the carboxylic acid is suppressed and the metabolite behaves as a basic drug, yielding good recovery by this procedure. The zwitterionic metabolite thus isolated can be alkylated during sample work up, often avoiding the need for any further derivatisation steps prior to GC or GC-MS analysis. Esterification occurs spontaneously during the washing step, methyl esters resulting from the use of methanol, ethyl esters from ethanol and CD_3, is incorporated from the use of deuterated methanol. We believe that there is a catalytic effect due to the silica which functions both as an acid and a dehydrating agent in the esterification reaction. No loss of the ester has been found during the passage of the alcohol, indicating that it is being retained by the ion exchange mechanism due to the basic part of the molecule. Elution of the esterified metabolite is achieved with 5 to 10% TEA in ethyl acetate.

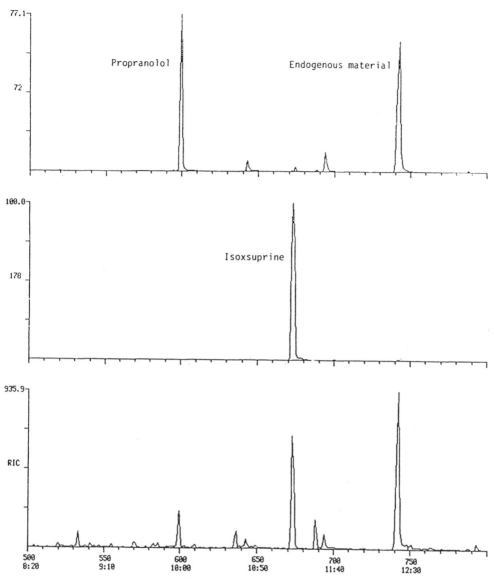

Figure 1. Top; Selected ion chromatogram for m/z 72
Middle; Selected ion chromatogram for m/z 178
Bottom; Reconstructed ion chromatogram

CAPILLARY GAS CHROMATOGRAPHY - MASS SPECTROMETRY (GC-MS) OF FENTANYL β-KETO ESTER

Blank urine samples were spiked with Fentanyl β-keto acid reference material and extracted by the above procedure. The extracts were analysed by combined GC-MS in the electron impact mode, scanning the mass range 40 to 450 amu in 1 second. An SE 54 capillary column, 7 m in length by 0.33 mm i.d. and 1 μm film thickness, was used for analysis. Split/splitless injections were made at 250° with an interface temperature of 310°. The

165

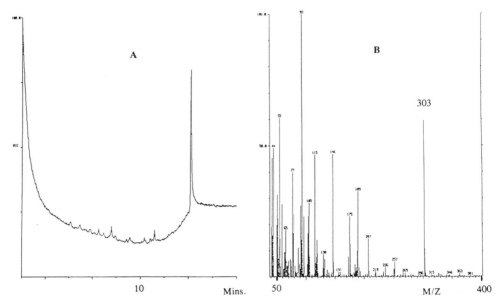

Figure 2.   GC-MS analysis of Fentanyl β-keto acid methyl ester.

column programme was isothermal at 75° for 1 minute then 18° per minute to 290°, then isothermal for 5 minutes. Results obtained by this procedure are shown in Figure 2. The previously published method of Frincke and Henderson [3] for the isolation of Fentanyl β-keto acid required 4 to 5 days to complete. In comparison samples can be analysed by this procedure in about 1 to 2 hours.

CONCLUSIONS

A stable elution reagent has been used successfully in the recovery of a wide variety of basic drugs isolated by SPE using phases with cation exchange character. Solid phase extraction has been shown to be of utility in the isolation of acidic metabolites of basic drugs where is it possible to carry out alkylation during the sample work up procedure. Extension of this work to post administration samples will follow.

REFERENCES

1.   M.C. Dumasia and E. Houghton. Screeing and confirmatory analysis of β-agonists and β-antagonists and their metabolites in horse urine by capillary gas chromatography-mass spectrometry. *J. Chromatogr. Biomed. Applns.*, 564, 503-515 (1991).
2.   L.M. Young and V. Robillo. A study of Fentanyl: Metabolism and excretion by the horse. *Proc. of the 6th Int. Conf. of Racing Analysts and Veterinarians*, Hong Kong, p257 (1985).
3.   J.M. Frincke and G.L. Henderson. The major metabolite of fentanyl in the horse. *Drug Metab. Dispos.*, 8, 425-427 (1980).

A NOTE ON ASPECTS OF SAMPLE PREPARATION FOR DRUG ANALYSIS IN SERUM BY CAPILLARY ELECTROPHORESIS

Helena Soini[1], Marja-Liisa Riekkola[2] and Milos V. Novotny[3]

[1] Orion Pharmaceutica
P.O. Box 65
SF-02101 Espoo
Finland

[2] Helsinki University
Dept. of Chemistry
Vuorikatu 20
SF-00100 Helsinki
Finland

[3] Indiana University
Dept of Chemistry
Bloomington
IN 47405
USA

INTRODUCTION

Drug analysis in a complex matrix like serum or plasma is one of the most challenging tasks for an analytical chemist. Knowledge of the drug molecule and possible matrix compounds as well as the analytical measurement technique all need to be considered when searching for a successful analytical quantification method. For the analysis of polar, water-soluble drug molecules, the analytical technique of choice has traditionally been high performance liquid chromatography (HPLC). However, in many instances the lack of separation efficiency restricts the use of HPLC. In biomedical analysis the requirements for an analytical technique are high resolution and selectivity of the separation system, especially when general detection, e.g. UV absorbance, is used. The compatibility of the injected solution with the analytical system is essential. Furthermore the analytical method should remain stable and precise over long analysis periods when a large number of samples are analysed.

Capillary electrophoresis (CE) is a separation technique with extremely high resolution capabilities [1,2]. Ionic species migrate in the applied electrical field according to the net effect of the electrophoresis mobilities and electroosmotic flow in the capillary filled with buffer solution. Terabe et al [3,4] have extended the technique to micellar electrokinetic capillary chromatography (MECC) by adding surfactants into the buffer. The presence of the pseudo-phase classifies MECC as a chromatographic technique. The use of MECC allows separation of neutral and ionic species in the same run.

So far capillary electrophoresis has been utilised relatively sparingly in quantitative drug analysis in biological samples. Here we discuss different sample preparation possibilities and topics which are important to produce CE compatible samples for quantitative drug analysis. Examples of liquid-liquid extraction in the analysis of naproxen and electrochromatographic solid-phase extraction [5] in the analysis of cimetidine, using UV absorbance detection, are presented.

Sample Preparation for Biomedical and Environmental Analysis,
Edited by D. Stevenson and I.D. Wilson, Plenum Press, New York, 1994

EXPERIMENTAL

All solvents and reagents used were analytical grade. The water was distilled and ion exchanged. Buffer solution were filtered through a 0.2μm filter (Sartorius GmbH, Göttingen, Germany).

The Waters Quanta 4000 capillary electrophoresis system (Waters Asociates, Milford, MA, USA) with UV detection at 254nm, was used for determination of naproxen. Ketoprofen was added to 1.0ml of serum as an internal standard and the sample was extracted with hexane-diethyl ether in acidic conditions. The organic layer was evaporated to dryness and dissolved in 50 to 200μl of water-ethanol (1:1). Automatic hydrodynamic injection was performed for 10 secs into the capillary (60cm x 75μm, i.d.). The sample was separated using +20kV with 10mM Trizma-10mM Tricine, pH8.0 buffer. A homemade CE system with UV detection at 229nm was used for determination of cimetidine. Injections were performed manually in the hydrodynamic mode. To 0.5ml of serum ranitidine was added as internal standard and the sample was transferred into a C18 solid-phase extraction cartridge. After washing, the sample was eluted with 50% tetrahydrofuran (THF) buffer using a voltage of 150V as the driving force as described elsewhere. Fractions of 20 to 50μl were collected and analysed by MECC (voltage 20kV). The capillary (60cm x 50μm, i.d.) was filled with a buffer consisting of 10mM HTAB - 9mM dihydrogen sodium phosphate at pH 6.4.

RESULTS AND DISCUSSION

The low injection volume in CE methods is a sensitivity-limiting factor. When injecting filtered serum or plasma directly into the MECC system, the effort of sample preparation is avoided. However, that approach is possible only for samples containing relatively high concentrations of drug.

Precipitating proteins with sufficient volumes of acid, acetonitrile or acetone [6] dilutes the sample further. Therefore, such methods are not recommended in connection with a concentration sensitive detector without any extra concentration steps. With high analyte concentrations such methods are suitable, producing disturbance-free backgrounds in electropherograms. In the determination of naproxen a liquid-liquid extraction method used for HPLC was transformed with minor modifications into CE. In Figure 1 electropherograms of a blank serum and a naproxen serum sample (5μg/ml) are shown. The precision of analysing samples of 10μg/ml and 1μg/ml was 1.6% and 1.3%, respectively (n=6). For the determination of cimetidine, an electrically-driven solid-phase extraction system was developed to minimise the eluent volume. As a secondary effect less background components were coeluted and a higher precision (4.5%, RSD, n=4) was achieved compared with the pressure-driven solid-phase extraction (18.9%, RSD, n=6) when 20μl sample volumes were collected.

CONCLUSIONS

Capillary electrophoresis is a highly selective, precise and fast analysis technique. It is suitable for the analysis of drugs in biological samples in CE and MECC modes. Basically the same sample preparation approaches as for HPLC and gas chromatography could be used with minor modifications. The current methods do, however, lack sensitivity in low analyte concentrations. It is therefore desirable to develop CE-compatible preconcentration methods. Examples could include the electrochromatographic solid-phase extraction methods would have improved detector design and the use of selective derivatisation techniques. The need for on-line sampling methods for quantitative analysis is also obvious. In future appli-

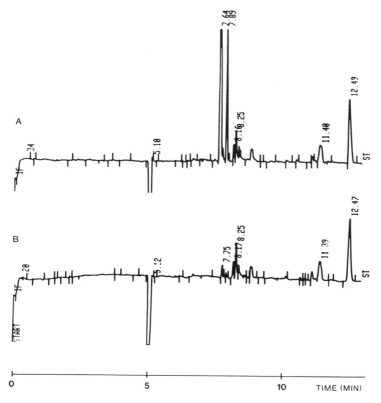

Figure 1. Electropherograms, (a) a serum sample extract containing naproxen 5µg/ml (7.89 min) and keto profen 25µg/ml (7.64 min) and (b) a blank serum extract.

cations of drug analysis in biological samples, capillary electrophoresis will be best for problems where HPLC lacks separation efficiency. These include especially the analysis of large analyte molecules and the separation of structurally similar metabolites and optical isomers.

REFERENCES

1. J.W. Jorgenson and K.D. Lukacs. Zone electrophoresis in open-tubular glass capillaries. *Anal. Chem.* 53, 1298-1302 (1981).
2. J.W. Jorgenson and K.D. Lukacs. Capillary zone electrophoresis. *Science* 222, 266-272 (1983).
3. S. Terabe, K. Otsuka, K. Ichikawa, A. Tsuchiya and T. Ando. Electrokinetic separations with micellar solutions and open-tubular capillaries. *Anal. Chem.* 56, 111-113 (1984).
4. S. Terabe. Electrokinetic chromatography: an interface between electrophoresis and chromatography. *Trends Anal. Chem.* 8, 129-134 (1989).
5. H. Soini, T. Tsuda and M.V. Novotny. *J. Chromatogr.* 3rd Int. Symp. on HPCE'91 Issue, in press.
6. J. Blanchard. Evaluation of the relative efficacy of various techniques for deproteinising plasma samples prior to high-performance liquid chromatographic analysis. *J. Chromatogr.* 226, 455-460 (1981).

STRATEGICAL CONSIDERATIONS IN APPROACHING SAMPLE PREPARATION

FOR PESTICIDE RESIDUE DETERMINATION

Philip J. Snowdon

Schering Agrochemicals Limited
Chesterford Park Research Station
Saffron Walden, Essex

SUMMARY

Hydrolysis, steam distillation, rapid bromination, sweep co-distillation and trace enrichment are all examples of valuable techniques which may be employed to achieve effective sample preparation in pesticide residue analysis. The principles and relative merits of each of these techniques are discussed within the context of specific sample preparation strategies. These cover a range of pesticide/substrate combinations, placing the emphasis upon strategical considerations imposed by the chemistry of the analyte(s), the nature of the matrix and the aims of the study, including the desirable limit of determination.

INTRODUCTION

In pesticide residue analysis, effective sample preparation is, together with representative sampling, fundamental in obtaining reliable and meaningful data. But what is residue analysis? For the purposes of this chapter, residue analysis concerns the trace analysis of pesticides and their metabolites in matrices of plant, soil and animal origin. The quantities involved are typically <0.1 mg/kg (ppm). Broadly speaking, the need to perform residue analysis arises within two distinct environments:-

1) Monitoring against prescribed tolerance levels. This often applies to regulatory laboratories or contract facilities assigned the task of screening specific commodities for residues of a wide range of commercial agrochemical products registered for use on the crops concerned. This type of work usually employs multi-residue methods (analysis for groups of compounds such as, for example, organochlorine insecticides or carbamates);

2) As part of product development within an agrochemical company. The requirement here is to develop residue methods for the determination of novel pesticides and their metabolites, with the aim of producing data in support of product registration. To achieve this, field trials conducted on a worldwide basis yield samples for analyses

Sample Preparation for Biomedical and Environmental Analysis,
Edited by D. Stevenson and I.D. Wilson, Plenum Press, New York, 1994

171

which are often specific to a particular compound. Method development related to this type of work is usually conducted in close liaison with metabolism chemists studying the fate of the compound in plant, soil, water and animal systems.

This paper focuses upon the second category, from which selected examples have been drawn to illustrate some different strategies employed for sample preparation in residue analysis. Due regard has been given to the factors which are considered when developing such strategies. The examples do not include the use of solid phase extraction (SPE) cartridges which, although successful in many residue analysis procedures, have been more than adequately covered elsewhere in this book, with numerous applications in the recent scientific literature.

The strategies covered make use of the following techniques:

> Acid hydrolysis
> Enzyme hydrolysis
> Steam distillation (liquid/liquid extraction)
> Rapid bromination
> Sweep co-distillation
> Valve switching/trace enrichment
> Extraction discs

Each will be illustrated with respect to particular pesticide/matrix combinations.

SAMPLE PREPARATION

This is sometimes mistakenly regarded as the final step in preparing a sample extract for determination by the chosen method, typically gas chromatography (GC) or high performance liquid chromatography (HPLC).

More correctly, sample preparation encompasses all of those steps which are necessary in preparing the original laboratory sample for determination by the chosen method, including the extraction process as well as subsequent clean-up of the extract. This is important because strategically the extraction method chosen can influence the entire sample preparation procedure by precluding certain approaches (see later).

GENERAL CONSIDERATIONS

What, then, are the key factors which must be taken into consideration when deciding a sample preparation strategy in residue analysis? Perhaps they may be categorised as follows:

> Chemistry of the analyte(s)
> Nature of the matrix
> Objectives of study
> Limit of determination

The relative importance of these will vary depending on the case in question.

R = H :   Clofentezine

R = OH :   4-hydroxyclofentezine

Structure 1. The acaricide clofentezine and its 4-hydroxy-metabolite.

STRATEGIES AND TECHNIQUES

Hydrolysis

There are two main reasons why hydrolysis may form part of sample preparation:

1) Metabolism studies show that in the substrate of interest, the pesticide and/or its metabolites exist mainly in conjugated form, i.e. as synthetic products formed with endogenous chemicals. These products are difficult to characterise and quantify, and analysis usually proceeds via acid or enzyme hydrolysis to release the pesticide/metabolite moiety;
2) The parent compound and its metabolites may hydrolyse to a common product which may be convenient for analytical purposes. Simple acid or base hydrolysis is normally sufficient for this, although more severe conditions are occasionally required.

The acaricide clofentezine (Structure 1) is used primarily on top fruit such as apples against the red spider mite (*Panonychus ulmi*), with particular activity as an ovicide. In the fruit itself, no significant metabolites occur and the analytical method developed measures parent compound only, using HPLC for final determination and incorporating a solid phase extraction clean-up. Many of the apples which are grown commercially, however, are subsequently pressed to obtain fruit juice, leaving the remaining solids or 'pomace' which may be incorporated into animal feeds. In such situations the determination of residues in meat and milk also becomes relevant.

In animals, clofentezine metabolism is more extensive. Most of the residue in meat and milk occurs as conjugated forms of the 4-hydroxy derivative (Structure 1).

These residues in milk are extracted intact by a simple procedure involving homogenisation with acetone and hexane and subsequent centrifugation. The top hexane layer contains the butterfat but negligible residue and therefore may be discarded. (It is worth considering that, if milk residues contained significant amounts of the 'free' parent molecule, which is non-polar in nature, such an approach could not be taken.) The residue remains in the aqueous acetone layer which is easily decanted from the solid protein pellet.

After removal of the acetone by rotary evaporation, the remaining aqueous extract is adjusted to pH 5 and digested at 50 °C for 2 hours in the presence of snail (*Helix pomatia*) digestive juice which contains both glucuronidase and sulphatase enzymes. This procedure is sufficient to release the 4-hydroxy-metabolite from its conjugated forms, the exact structures of which are unknown. Acidification then allows partition of the metabolite residue into petroleum ether prior to concentration and final determination by HPLC.

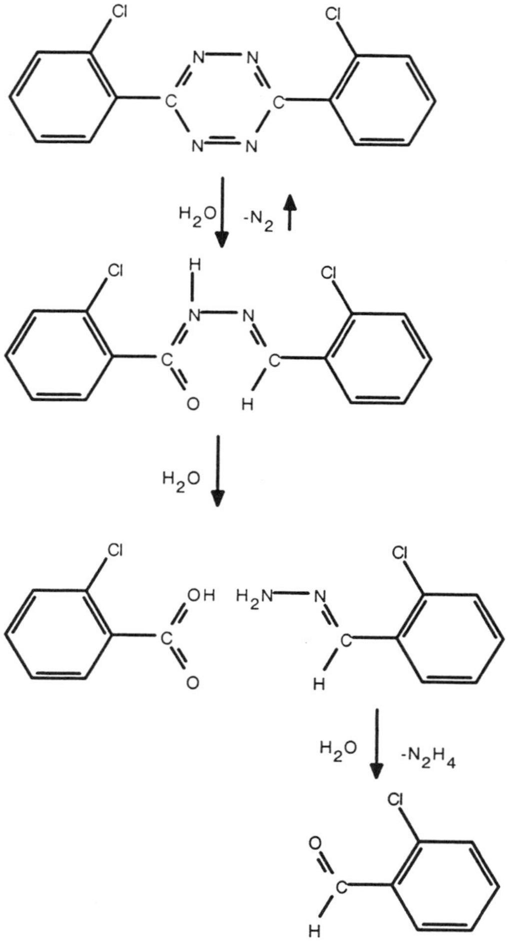

Scheme 1. Postulated mechanism for the acid hydrolysis of clofentezine.

In meat, conjugated 4-hydroxy-clofentezine residues are not readily extractable and thus a different strategy is required. The favoured approach is to directly hydrolyse the tissue samples by heating under reflux with concentrated hydrobromic acid. This process breaks the central tetrazine ring (Scheme 1) and releases 1 mole of 2-chlorobenzoic acid (2-CBA) from each mole of free or conjugated 4-hydroxy-clofentezine (1). The stoichiometry of this reaction was investigated and validated through fortification tests with various hydroxylated derivatives and by analysis of samples containing ^{14}C radiolabelled residues.

This hydrolysis strategy takes advantage of the chemistry of the residues' tetrazine moiety and provides satisfactory extraction efficiency whilst yielding a common product, 2-CBA, which is readily removed from the tissue hydrolysate by partition into diethyl ether. After further clean-up (by partition into alkali then back into diethyl ether), the 2-CBA is methylated to facilitate GC determination using mass selective detection in selected ion mode (allowing confirmation of residues by characteristic fragment ion ratios).

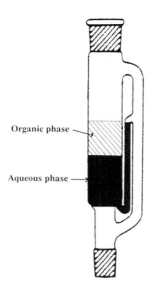

Organic phase

Aqueous phase

Figure 1.   Liquid/liquid extractor.

Steam Distillation

Steam distillation follows on logically from hydrolysis, as the two are often applied simultaneously. Steam distillation using liquid/liquid extractors is both a simple and very effective extract clean-up technique which, by definition, is restricted to use with steam volatile compounds. The chemistry of a pesticide and its metabolites is therefore the prime consideration in deciding the feasibility of this approach, but, because of the efficiency of clean-up offered (extracts are often suitable for final chromatographic determination), it is worthwhile examining possibilities for deliberately producing steam volatile derivatives within the residue method, hence the association with hydrolysis.

To operate this principle, the liquid/liquid extractor (Figure 1) is connected between a round-bottomed flask, containing an aqueous extract, and a water-cooled condenser. The extractor itself is primed with water and a suitable organic solvent of lower density, typically hexane or petroleum ether. Upon reflux, steam volatiles distil over with the water vapour and the condensates are extracted into the upper, organic phase whilst the extracted water is returned to the reaction flask by a continuous syphoning mechanism. A teflon baffle is sometimes added to the extractor to aid mixing during extraction.

Amitraz (Scheme 2) is a triazapentadiene type insecticide with uses which range from parasiticidal activity in animal health products to acaricidal activity for use on crops such as citrus fruit, apples and hops. Its metabolic degradation in plants (Scheme 2) yields two major metabolites (BTS 27271 and BTS 27919) which retain the common 2,4-dimethylaniline (2,4-DMA) moiety.

The strategy for this compound on all crop substrates thus involves deliberate hydrolysis of the parent compound and metabolites through to 2,4-DMA itself, which is steam volatile. Once again, looking at the stoichiometry of the reactions involved (Scheme 2), it may be seen that 1 mole of amitraz gives rise to 2 moles of 2,4-DMA. By multiplication of the final 2,4-DMA analysis result with the appropriate molecular weight factor, the total residue of the combined parent compound and its metabolites is expressed in terms of the most toxic component, namely the parent itself, thereby meeting the requirements of regulatory authorities and the objective of the residue study.

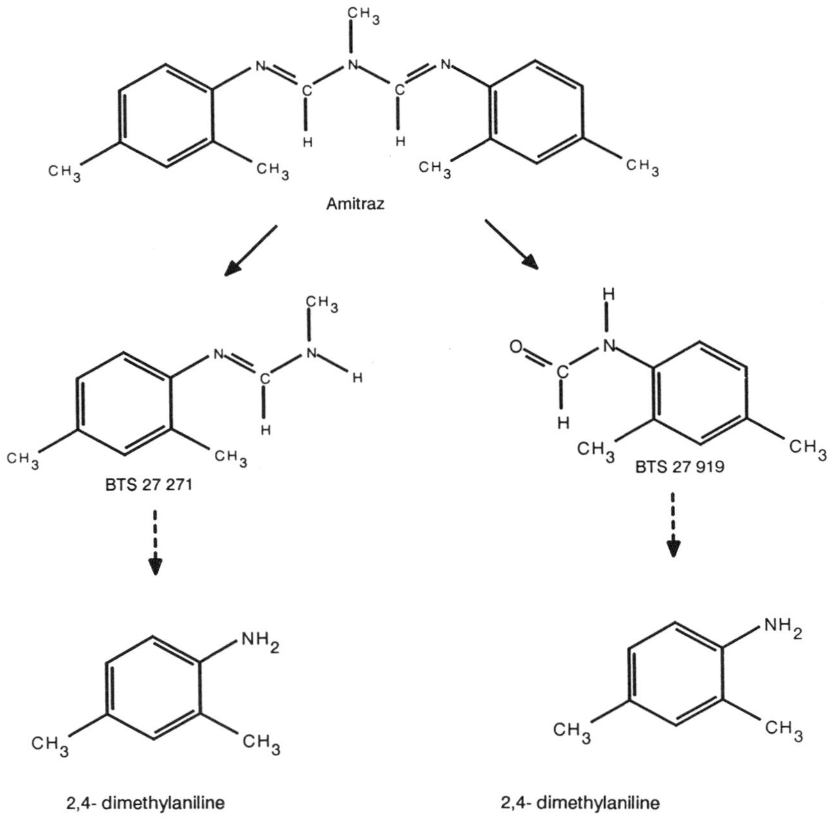

Scheme 2.   Plant metabolism of amitraz.

Amitraz itself readily hydrolyses under acidic conditions to give BTS 27271 and BTS 27919 (the major plant metabolites), which in turn hydrolyse to 2,4-DMA more rapidly under alkaline conditions. Hydrolysis is therefore conducted in two stages: reflux in acid, to hydrolyse amitraz (this also breaks conjugated residues), followed by alkaline hydrolysis with simultaneous liquid-liquid extraction of 2,4-DMA as it is formed. The alkaline conditions of this second hydrolysis promote the steam distillation of the aniline by suppressing its weakly basic nature. Final determination by GC with electron capture detection (ECD) follows after derivatisation of the aniline (Figure 2).

A similar strategy is suited to prochloraz (Scheme 3). This compound is an imidazole fungicide widely used on cereals and oilseed rape against *Ascomycetes* and *Fungi imperfecti* by inhibiting ergosterol biosynthesis. Prochloraz is also effective as a post-harvest treatment against storage diseases in various tropical fruits.

The plant metabolism of prochloraz is shown in Scheme 3, with free and conjugated forms of the metabolites BTS 44596 and BTS 44595 comprising the major part of crop residues. In this case, all components of the residue contain the common 2,4,6-trichlorophenoxy moiety and, with the potential advantage of an extremely effective extract clean-up coupled with an easy route to the measurement of conjugated residues, hydrolysis of all components to 2,4,6-trichlorophenol (2,4,6-TCP) is preferred. A complication, however, is that this is not readily achievable by simple acid or alkaline hydrolysis. Reflux with pyridine hydrochloride at 210 °C has proven to be the most reliable means of achieving quantitative yields. The resulting hydrolysate, charred in appearance, is readily washed into a round-

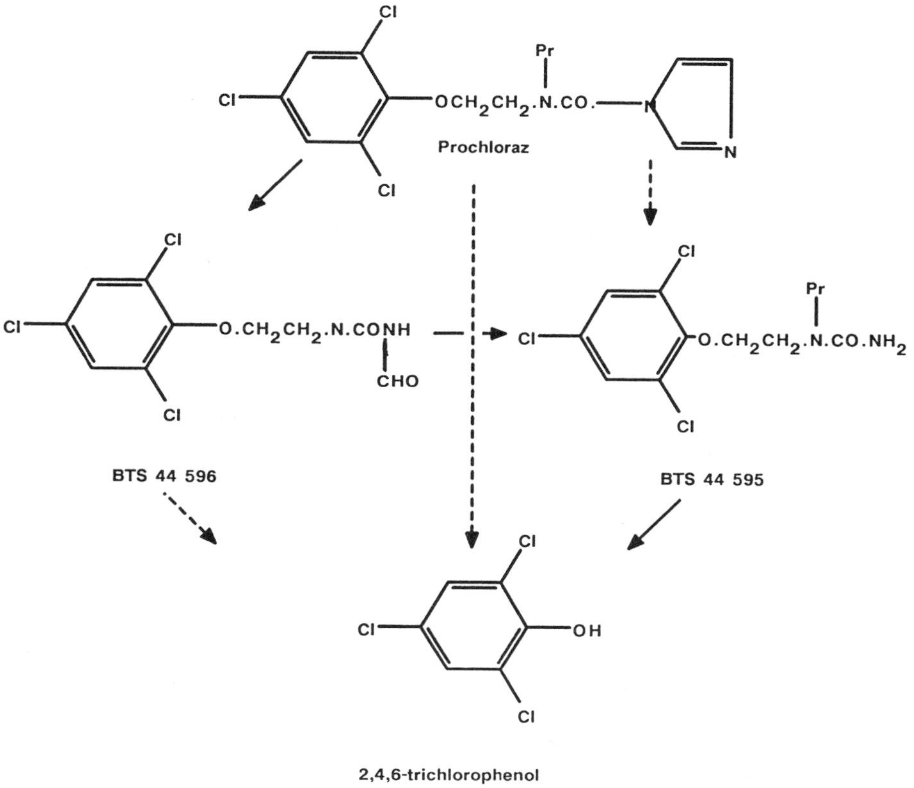

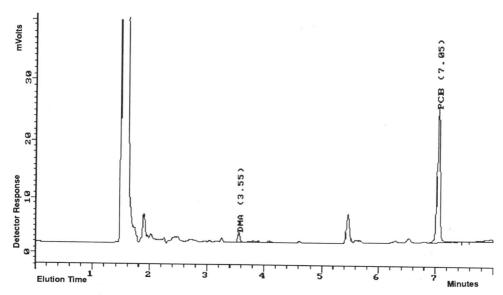

Scheme 3.   Plant metabolism of the fungicide prochloraz.

Figure 2.   Amitraz residue analysis. This typical chromatogram represents 84% recovery efficiency from blackcurrants fortified at 0.025 mg/kg. Analytical conditions were as follows:- Column: SPB 5, 20m x 0.32mm i.d., 0.25μm df; Carrier gas: Nitrogen at 1.7ml/min; Injection: Split, ratio 40:1, 250°C; Temp. prog: 140°C for 5 mins., then 10°C/min. to 170°C; Detection: ECD, 275°C.

bottomed flask with dilute acid to facilitate the subsequent steam distillation of the acidic 2,4,6-TCP. The final extract is suitable for direct determination by GC with ECD.

Rapid Bromination

As stated earlier, certain pesticides may yield substituted anilines which require a final derivatisation step to facilitate GC determination with ECD. Bromination is a common technique employed for this purpose, and features in residue methods for the determination of the fungicide carboxin, herbicides phenmedipham and desmedipham, the acaricide formetanate and thidiazuron, a plant growth regulator (Structures 2-6). Of these compounds, desmedipham and thidiazuron hydrolyse to aniline itself. This can lead to problems in certain plant matrices such as cereal grain where determination is restricted by naturally occurring components derived from the amino acid tryptophan. Under conditions which had become established as the accepted standard for bromination, employing nascent bromine from potassium bromate and potassium bromide in aqueous solution (2), tryptophan undergoes the competing reaction shown in Scheme 4. This leads to unsatisfactory apparent residues in untreated control samples with a consequent need to set undesirably high limits of determination.

One solution to this problem is to modify the sample preparation procedure to include a more stringent clean-up prior to derivatisation. A simple alternative, however, is offered by adjustment of the bromination conditions in accordance with the findings from a study of the reaction kinetics (3). The rate determining step in the formation of 2,4,6-tribromoaniline (TBA) from tryptophan is the further bromination of 2-amino-3,5-dibromoacetophenone (ADBAP) to TBA in Scheme 4.

Essentially, if the bromination of the pesticide-derived aniline can be effected very rapidly, the background contribution of TBA from tryptophan is minimised. In practice, such control may be achieved by using a bromine water/hydrobromic acid system which produces quantitative bromination of the aniline in five seconds, after which time the reaction can be terminated by addition of saturated sodium sulphite solution. This alternative has been successfully validated by recovery efficiency testing, with examples of typical data from tests with carboxin and its sulphoxide metabolite presented in Table 1.

Table 1.   Carboxin recovery efficiency data

	Cereal grain	Straw	Immature plant	Soil leachate
Fortification range (mg/kg)	0.02-2.0	0.05-0.5	0.025-0.25	0.01-0.02
No. of tests	20	11	10	3
Mean recovery (%)	90	99	92	96
CV (%)	11	7.1	10	7.1
LOD (mg/kg)	0.01	0.05	0.02	0.01

2

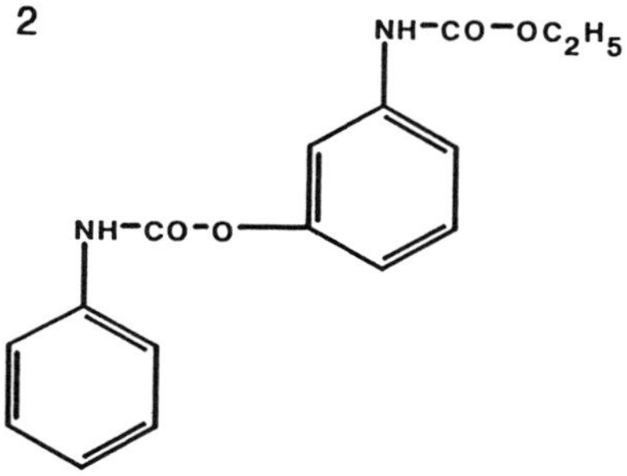
NH-CO-OC2H5
NH-CO-O

Desmedipham

3

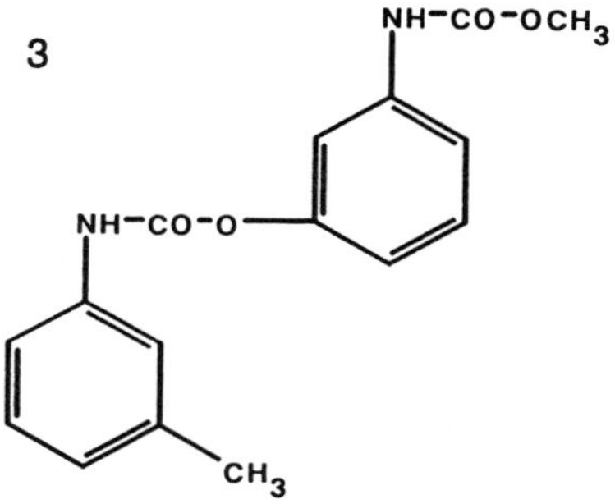
NH-CO-OCH3
NH-CO-O
CH3

Phenmedipham

4

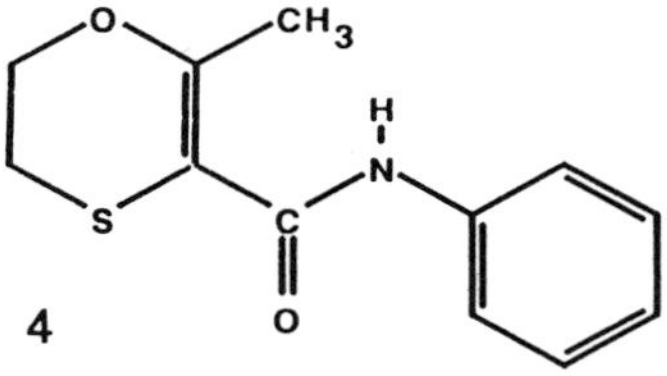
O
CH3
S
H
N
C
O

Carboxin

5

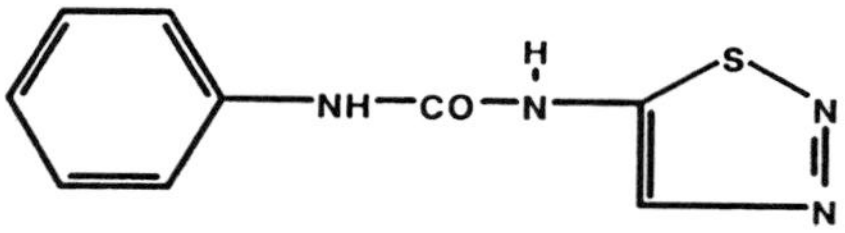
H
NH-CO-N
S
N
N

Thiadiazuron

6

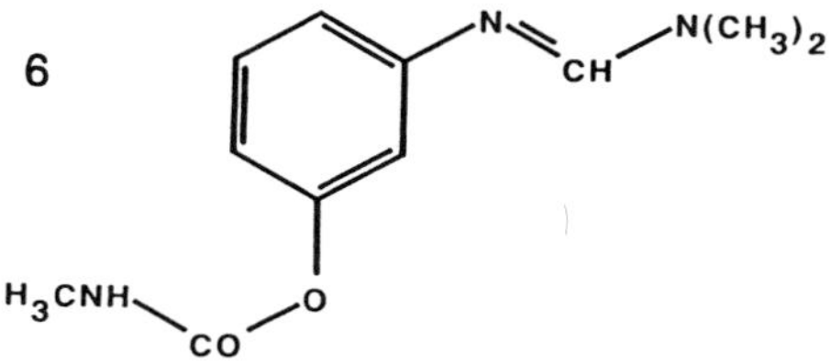
N
CH
N(CH3)2
H3CNH
CO
O

Formetanate

Structures 2-6. Examples of pesticides determined via bromination of substituted aniline moieties.

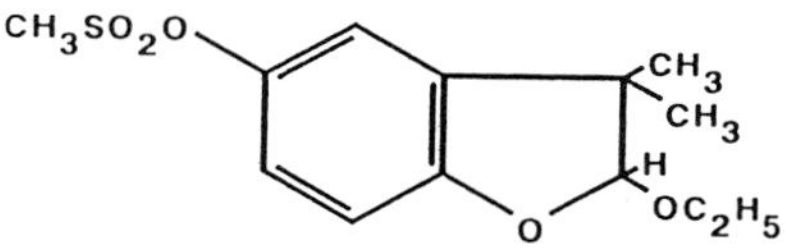
CH3SO2O
CH3
CH3
H
OC2H5
O

Structure 7. The herbicide ethofumesate.

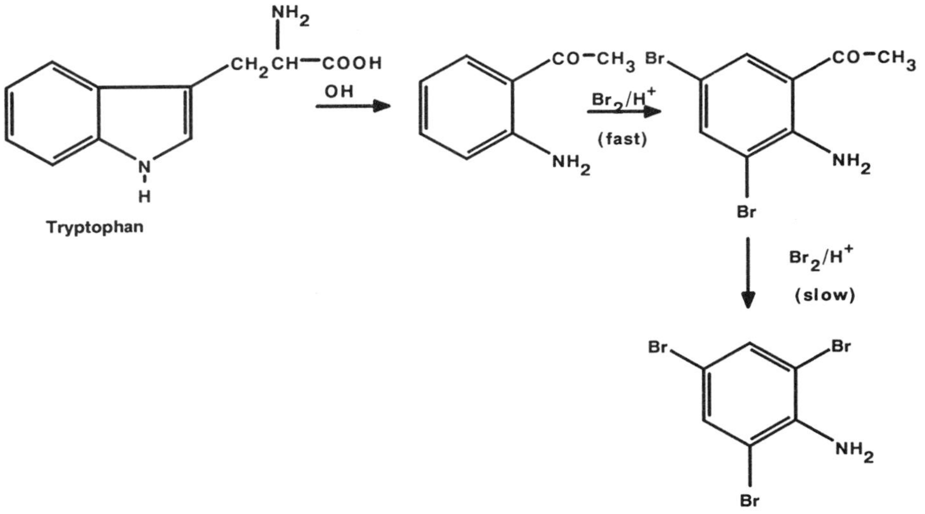

Scheme 4. Bromination of tryptophan.

Sweep Co-Distillation

Sweep co-distillation is a sample preparation technique developed some 25 years ago (4) for the clean-up of extracts of semi-volatile organochlorine and organophosphorus pesticides in matrices of an oily or fatty nature. Despite the apparent unpopularity of this method in recent years, there have been developments to simplify the handling of the rather cumbersome equipment involved, and modern systems remain very effective for appropriate compounds in matrices such as vegetable oils and dairy products. The nature of the matrix and the volatility of the analytes are thus the prime strategical considerations in selecting this technique.

The herbicide ethofumesate (Structure 7) is used as a pre- and post- emergence treatment against broad-leaf and grass weeds in beet crops and pasture grasses, but has minor uses in crops such as tobacco and camomile which leave viscous oily residues that are difficult to handle by conventional solid phase extraction methods. Using sweep co-distillation (Figure 3), concentrated extracts equivalent to 1-2 g of crop may be diluted in pure corn oil and injected into the fractionation tube at an oven temperature of 230 °C and a nitrogen carrier gas flow of approximately 200 ml/min. Under these conditions, the oil disperses as a thin film on the surface of the glass beads inside the annular space within the fractionation tube, allowing ethofumesate to partition into the gas stream before being subsequently trapped on a small column of Florisil. This process takes 35 to 40 minutes to complete, and up to ten extracts may be handled simultaneously. The final extract, eluted from the Florisil trap (which by adaptation, may be replaced by commercially available SPE cartridges) in a mixture of 4% ethyl acetate in hexane, is suitably free of non-volatile contaminants for final GC determination using flame photometric detection (FPD) in the sulphur mode.

Trace Enrichment

When working with water samples, a major strategical consideration in the European Community is the desirable limit of determination, particularly when complying with the drinking water directive requiring a determination limit of 0.1 ppb for individual pesticides,

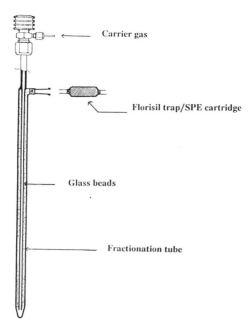

Carrier gas

Florisil trap/SPE cartridge

Glass beads

Fractionation tube

Figure 3.   Sweep co-distillation.

irrespective of their relative toxicities. Such situations require that sample preparation involves a significant concentration of residues to facilitate final detection and measurement. Using solid phase extraction cartridges packed with reversed-phase material allows trace levels of relatively non-polar compounds to be extracted and concentrated from several litres of water, if necessary, and subsequently eluted with as little as 1 to 2 ml of a less polar solvent. Passing large volumes of water through these cartridges can, however, be very time consuming, particularly when several samples are to be analysed. A variation on the theme which is faster and easier to use in practice, especially for batch analysis, is the extraction disc.

Extraction discs consist of chemically bonded silica particles enmeshed in an inert PTFE matrix to form a mechanically stable disc 0.5 mm thick, with a typical diameter of 47 mm, designed to fit standard vacuum filter holders. Large volume water samples may then be drawn through the disc under vacuum at flow rates of 20 to 30 ml/min. After vacuum drying, the residue may be eluted with a few millilitres of a suitable organic solvent, thereby giving a large concentration factor. This method is currently being adopted in our laboratories as the first approach to the determination of residues in drinking water. It applies mainly to newer development compounds, although some more established products have been re-evaluated by this technique. Performance at the 0.1 ppb level continues to be encouraging, with recovery efficiencies typically between 80 and 90%.

In certain cases where an analyte may be conveniently determined by HPLC, direct on-line sample preparation may provide successful trace enrichment of water residues via an HPLC pre-column. As the entire extract effectively reaches the HPLC detector, sample volumes as low as 20 to 50 ml may suffice to attain the desired limit of determination.

In a system of the type shown in Figure 4, water samples containing residues of the herbicide ethofumesate (referred to earlier), may be loaded onto a 10 mm x 2.1 mm i.d. pre-column of C18 reversed-phase material at a rate of 5 ml/min. A sample size of 20 ml pre-

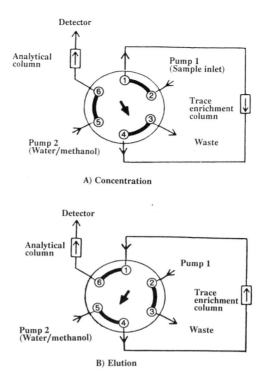

A) Concentration

B) Elution

Figure 4.   HPLC valve-switching configuration for on-line trace enrichment of water residues.

cludes the risk of breakthrough.   After trace enrichment, switching the six-port valve position for one minute is sufficient to backflush the residue from the pre-column onto an analytical C18 column (200 mm x 3.0 mm i.d.), eluting with 68% methanol in water. Final detection is by UV absorption at 277 nm.

CONCLUSIONS

With reference to specific examples, an attempt has been made to illustrate some different strategical approaches to sample preparation in pesticide residue analysis using hydrolysis, steam distillation, rapid bromination, sweep co-distillation and trace enrichment. In all cases, the chemistry of the pesticide being determined, the nature of the matrix concerned and the aims of the study, including the desired limit of determination, are important considerations in deciding the strategy to develop.

REFERENCES

1.   P.J. Snowdon, R.J. Whiteoak and J.D. Manley. The hydrolysis of clofentezine and related tetrazines as the basis of determination of residues in bovine tissues. *Fresenius J. Analytical Chemistry*, 339 : 444 (1991).
2.   N.A. Jenny, K. Kossmann and Formetanate. **In:** Analytical Methods For Pesticides and Plant Growth Regulators, Vol. VII, p. 290, G. Zweig ed. Academic Press (1973).
3.   J.D. Manley, P.J. Snowdon and R.J. Whiteoak. Poster 08B-41, IUPAC 7th Int. Congress of Pesticide Chemistry, Hamburg (1990).
4.   R.W. Storherr and R.R. Watts. Sweep co-distillation clean-up method for organophosphate pesticides. I. Recoveries from fortified crops, *JAOAC*, 48 : 1154 (1965).

A NOTE ON THE DETERMINATION OF ORGANOCHLORINE PESTICIDES IN

WATER

T. Prapamontol and D. Stevenson

Robens Institute of Health and Safety
University of Surrey
Guildford, Surrey
GU2 5XH. UK

INTRODUCTION

The effectiveness and potential economic benefits of pesticides have led to their widespread use in modern agriculture. Despite their undoubted success in increasing food production an increasing number of the public perceive their use as unnecessary and harmful to both man the environment. As such a balance between the beneficial and adverse effects of pesticides is sought and analytical methods that are sensitive, reliable and robust are required [1,2].

Many organochlorine pesticides (OCPs) which have shown undesirable effects in man and the environment have been banned in developed countries, and replaced by less persistent pesticides. Unfortunately, the new products are not always accessible to developing countries and a number of OCPs such as dichlorodiphenyltrichloromethane (DDT), dieldrin, aldrin, endrin, aned chlordane are still in use. This, in combination with their persistence in the environment means that analytical methods are required both for developed and developing countries. Concern about the quality of water supplies and possible contamination by pesticides has resulted in intensified monitoring programmes [3,4,5]. Environmental monitoring also requires the analysis of foodstuffs, crops, soil and biological samples such as blood, urine and breast milk [6].

In recent years advances in analytical instruments have enabled us to measure pesticide residues in biological and environmental samples down to nanogram and picogram levels. This is partly due to the availability of highly sensitive and specific detection (primarily the electron capture detector in the case of OCPs). The introduction of high resolution fused silica capillary columns has allowed efficient multicomponent separations of OCPs. However the very low levels required and the complex matrices encountered means that sample preparation is the rate determining step for this type of analysis. The work described here has explored and optimised the use of solid phase extraction cartridges for extracting and concentrating water samples and has explored feasibility of using these for on-site collection. The method was modified to allow the determination of OCPs in breast milk.

Sample Preparation for Biomedical and Environmental Analysis,
Edited by D. Stevenson and I.D. Wilson, Plenum Press, New York, 1994

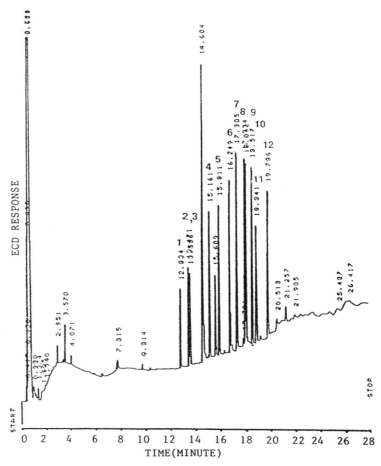

Figure 1.   OCP standards on a 10m x 0.32mm (i.d.) ’μm SE 52/4 column. Operating conditions: column oven temperature was programmed from 80°C (held for 1 min) with a 10°C/min ramp rate to 250°C and held for 10 mins, injection port temperature at 200°C, ECD detector oven at 320°C, helium carrier gas was set at 40cm/second liner velocity, OFN make-up gas for ECD was 60ml/min and splitless injection of 1μl pesticide solution. Peak identification: 1. α-HCH; 2. β-HCH; 3. ɤ-HCH; 4. Heptachlor; 5. Aldrin; 6. Hept. epoxide; 7. o,p’-DDE; 8. p,p’-DDE; 9. Dieldrin; 10. Endrin; 11. p,p’-DDD; 12. p,p’-DDT.

EXPERIMENTAL

OCP standards were all obtained from Promochem, St Albans. Solid phase extraction cartridges (C8 and C18 bonded silica, 200 and 500 mg) and a vacuum manifold from Jones Chromatography, Hengoed. Organic solvents (pesticide residue grade) were from BDH, Poole. The capillary gas chromatograph was a model 5890A from Hewlett-Packard, Wokingham. Capillary columns (10M x 0.32mm I.D.) 1μm SE 52/4 were from SAC Chromatography, Cambridge. Chromatographic conditions were: injection port 200°C, detector oven (ECD) 320°C, column oven temperature programmed from 80°C (held for 1 min) increased by 10°C/min to 250°C and held for 10 mins; helium carrier gas inlet pressure 4.3 p.s.i. which was 40 cm/s gas velocity, splitless injection of 1 μl.

RESULTS AND DISCUSSION

Successful separation of 12 common OCPs was achieved on a 10 M fused silica column (see Figure 1). Splitless injection with solvent effect was used for the introduction of the sample onto the capillary column. Three non-polar solvents, n-hexane (b.pt 68°C) ethyl acetate (b. pt. 77°C) and isooctane (b.pt. 99°C) were tried. They gave similar results so isooctane was chosen since its boiling point is higher. The last eluting component p,p'-DDT had a retention time of 19.8 mins giving a total run time of 30 mins to allow for cooling and re-equilibration. Using the column under these conditions gave > 50,000 theoretical plates.

The chromatographic conditions described above were used to optimise the extraction and pre-concentration of OCPs from water at concentrations down to 0.1 µg/L. A chromatogram of OCPs spiked into tap water is shown in Figure 2.

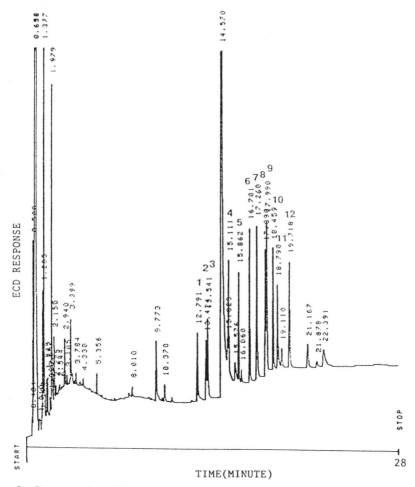

Figure 2. Representative capillary chromatogram of extracted OCP standard from spiked tap water. Operating conditions were the same as in Figure 1.

In order to maximise recovery it was found necessary to add a small volume (5-10%) of methanol to water samples. Cartridges were conditioned using 2 x 1 ml isooctane, ethyl acetate, methanol and distilled water. Spiked water samples were used to determine the volume breakthrough of cartridges (200 and 500 mg). At least 300 ml of water spiked with 0.02 to 0.05 µg/L of each OCP could be passed through the 200 mg cartridge and at least 400 ml through the 500 mg cartridge. Flow rates were 8-10 ml/min. In all cases the spiked water samples contained 5% methanol and OCPs were eluted from the columns using 2 x

Table 1. Comparison of volume breakthrough and sorbent capacity on C8 bonded silica cartridges.

OCPs	Conc in water (ppb)	C8 (200mg) Breakthrough (ml)	C8 (200mg) Capacity (ng)	C8 (500mg) Breakthrough (ml)	C8 (500mg) Capacity (ng)
α-HCH	0.01	400	4	1000	10
β-HCH	0.02	300	6	1000	20
ɣ-HCH	0.01	300	3	1000	10
Heptachlor	0.02	400	8	400	8
Aldrin	0.02	500	10	1000	20
Hept epoxide	0.02	ND[a]	ND[a]	400	8
o,p'-DDE	(0.20)	ND[b]	ND[b]	ND[b]	ND[b]
p,p'-DDE	0.03	400	12	400	12
Dieldrin	0.03	ND[a]	ND[a]	400	12
Endrin	0.04	500	20	500	20
p,p'-DDD	0.03	500	15	1000	30
p,p'-DDT	0.05	500	25	1000	50

Table 2. Recovery of OCPs from distilled and tap water using C8 cartridge extraction of 100ml spiked water (in duplicate).

OCPs	Spiked water (ppb)	% Recovery Distilled water	% Recovery Tap water
α-HCH	0.02	100	100
β-HCH	0.04	110	110
ɣ-HCH	0.02	120	120
Heptachlor	0.04	90	100
Hept epoxide	0.04	90	100
o,p'-DDE	(0.20)	(100)	(100)
p,p'-DDE	0.06	93	93
Dieldrin	0.06	133	133
Endrin	0.08	105	110
p,p'-DDD	0.06	87	93
p,p'-DDT	0.10	92	96

Key

ND[a] = not determined in the initial experiment
ND[b] = not determined since it was used as internal standard.

0.5 ml of isooctane. The results are shown in Table 1. Using 100 ml water samples on the 200 mg cartridges good recoveries and reproductibility were obtained for all OCPs tested, as shown in Table 2.

Linear standard curves were obtained for all compounds representative calibration plots for heptachlor epoxide and α-HCH are shown in Figure 3. The limit of detection for each OCP was at least down to 0.05 µg/L.

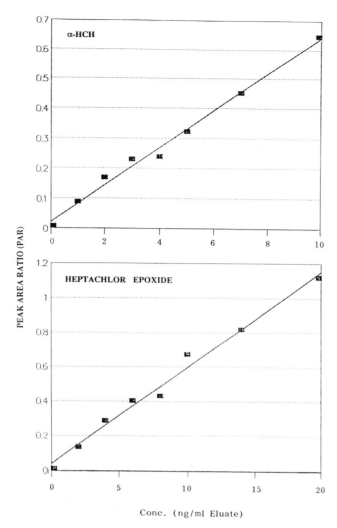

Figure 3.   Representative calibration curves of α-HCH and heptachlor epoxide extracted from spiked tap water according to the present method.

Environmental monitoring programmes for water often have to deal with a large number of bulky samples which need to be transported to analytical laboratories. Preliminary experiments to assess the suitability of solid phase extraction cartridges for storing extracted OCPs were carried out. A comparison between spiked tap water samples stored in glass bottles and C8 cartridges was made. The results (shown in Figure 4 and 5) suggest that for most of the OCPs storage on C8 cartridges gave higher recoveries than storage in bottles, with the possible exception of dieldrin after 12 days.

CONCLUSIONS

The simple method described here can determine all twelve OCPs down to at least the EEC drinking water recommended level (0.1 µg/L). The method is rapid with the cartridges providing both trace enrichment and clean up in one step. Solid phase extraction cartridges also show great potential for use on-site such that large volumes of water need not be transported to analytical laboratories.

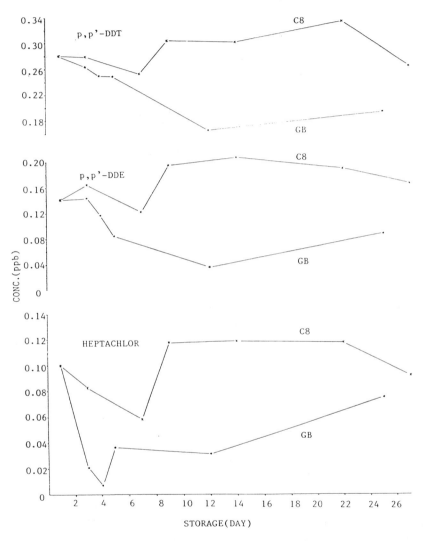

Figure 4. The effect of storage on p,p'-DDT; p,p'-DDE and hepatochlor when stored as water extracts on C8 cartridges (C8) and water in glass bottles (GB).

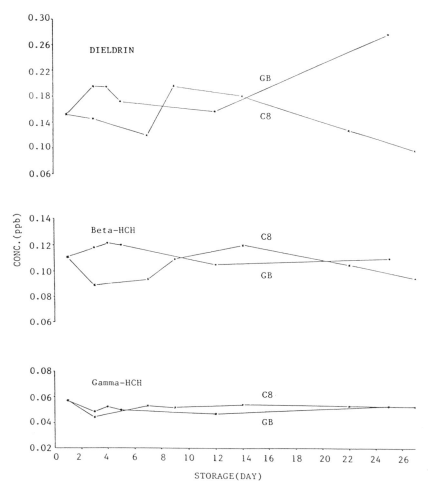

Figure 5.   The effect of storage on dieldrin, β-HCH and ɤ-HCH when stored as water extracts on C8 cartridges (C8) and water in glass bottles (GB).

REFERENCES

1.   G.A.Junk and J.J.Richard. Organics in Water: Solid phase extraction on a small scale. *Anal. Chem.*, 60, 451-454 (1988).
2.   M.J.M.Wells and J.L.Michael. Recovery of picloram and 2,4-D from aqueous samples by reversed-phase solid phase extraction. *Anal. Chem.*, 59, 1739-1742 (1987).
3.   E.Davoli, E.Benfenati, R.Bagnati and R.Fanelli. Analysis of atrazine in underground waters at part-per-trillion levels as on early warning method for contamination and for soil degradation studies. *Chemosphere*, 16, 1425-1430 (1987).
4.   M.J.M.Wells and J.L.Michael. Reversed-phase solid phase extraction for aqueous environmental sample preparation in herbicide residue studies. *J. Chromatogr. Sci.*, 25, 345-350 (1987).
5.   R.L.Nash. Solid phase extraction of carbofuran, atrazine, simazine, alachlor and ganazine from shallow well water. *J. Assoc. Anal. Chem.*, 73, 438-439 (1990).
6.   T.Prapamontal and D.Stevenson. Rapid method for the determination of organochlorine pesticides in milk. *J. Chromatogr.*, 552, 249-257 (1991).

ANALYSIS OF PESTICIDES AND PCB RESIDUES IN WATER, SOIL AND PLANT MATERIAL

W.P. Cochrane, D. Chaput and J. Singh

Agriculture Canada
Laboratory Services Division
Bldg. 22, C.E.F.
Ottawa, Ontario
K1A 0C6. Canada

SUMMARY

Pesticides are widely used in Agriculture and are divided into distinct classes, namely organochlorines (OCs), organophosphates (OPs), carbamates, herbicides and fungicides, etc. The OC, OP and carbamates exhibit a wide degree of persistence in the environment. This presents different health and safety concerns with the OCs being the most persistent and bioaccumulative while the OPs and carbamates present a more acute toxicity and less persistence.

The problem in the extraction and sample preparation of the non-polar OC residues is compounded by their co-extraction with the ubiquitous polychlorinated biphenyls (PCBs) which puts rigorous demands on the methods of sample clean-up. In the OP and carbamate areas, where one is dealing with the more highly polar substances, the extraction and clean-up solvents must be of similar polarity. In water analysis, emphasis must be placed on the partition coefficient of the residue between water and the extracting solvent, while with soil and plants, the solubility is a paramount consideration. A variety of adsorbants, e.g., silica, modified aluminas, etc. have been used in a variety of ways to eliminate interfering co-extractives. Successful identification and quantification of pesticides and PCB residues has relied heavily on gas chromatography with element-specific detectors including mass spectrometry.

Specific examples are provided from current laboratory operations.

INTRODUCTION

Currently, there are over 600 chemical pesticides registered for agricultural use. They can be classified in a number of fashions, however, one of the most accepted means is by chemical structure. Thus, we commonly refer to the organochlorines (OCs), the organophosphates (OPs), the carbamates, the triazines, the sulfonyl ureas, the phenoxy acids, the phenols, etc.

Sample Preparation for Biomedical and Environmental Analysis,
Edited by D. Stevenson and I.D. Wilson, Plenum Press, New York, 1994

During the last three decades, the analysis of agricultural and environmental substrates for pesticide residues has relied upon:

- single or multi-residue extraction/clean-up procedures;
- gas chromatography (GC) or high-pressure liquid chromatography (HPLC) with various element-specific detectors.

This approach has conveniently covered most of the thermally-stable lipophilic pesticides and some hydrophillic insecticides in a wide-range of sample types including water, soil, and agricultural crops, fruits, vegetables and meats. Thermally labile pesticides can be assayed by gas chromatography depending on the compound and GC conditions - but HPLC is more common (certain carbamates). Although most available residue methods have been developed for the detection and quantification of only a single pesticide or pesticide class, it is estimated that over 325 pesticide residues/metabolites can now be analysed directly, or after derivatisation, by GC using several specific detectors.

In recent years, the issue of pesticide residues has been highly visible for both the scientific community and the general public due to:

- environmental concerns related to PCBs, dioxins, etc.;
- food issues such as the 1989 Alar in apple and apple juice issue in the USA;
- increased governmental requirements (EEC, USA/Canada Free Trade) and;
- increased public awareness due to access to information legislation.

The public's fear of pesticide continues to grow, even though many reports have been issued to corroborate the safe use of pesticides and the safety of our food supply. In Agriculture Canada, this heightened public concern has manifest itself in a 10-fold increase in sample volumes from 13,205 to over 140,000 in the 6-year period 1985-1991 for chemical residues of pesticides, heavy metals, etc. in food commodities and environmental substrates.

MULTI-RESIDUE ANALYSIS - PLANTS AND SOILS

To handle the increased analytical demand one approach that we have taken over the years in the analysis of pesticide residues is to expand and automate a multi-residue method that covers a broad spectrum of chemical types and matrices. The non-polar OCs and intermediate polarity OPs have traditionally been assayed by the multi-residue procedure of Mills [1,2] using hexane or acetonitrile as extraction solvent followed by alumina/florisil clean-up and quantification by GC/electron capture detector (GC/ECD) for OCs or GC/flame photometric detector (GC/FPD) for OPs. The methodology is limited in that very polar water soluble pesticides, such as some OPs, may be lost in the extraction phase, while certain N-methyl carbamates degrade on florisil or the GC column. To overcome this latter problem, carbamate pesticides were analysed separately using the 1980 Krause procedure [3] which involves extraction with methanol, clean-up on a Nuchar-Celite column, separation by reversed-phase HPLC and fluorescence detection after, post-column derivatisation. This was simplified into a fast and efficient multi-residue screening method for N-methyl carbamates in fresh fruits and vegetables [4] which was subsequently incorporated into an expanded multi-residue method for the determination of OP, OC and carbamate residues in fruits, vegetables and soils. It involved extraction with methanol partitioning into methylene chloride and a schematic for handling clean vs. dirty extracts is shown (Scheme 1). Co-extractives are separated by gel permeation chromatography (GPC) [2] or GPC with on-line Nuchar-Celite cleanup for crops with high chlorophyll and/or carotene content (eg., cabbage and broccoli). The use of the GPC afforded an automated 23 sample cleanup step using a

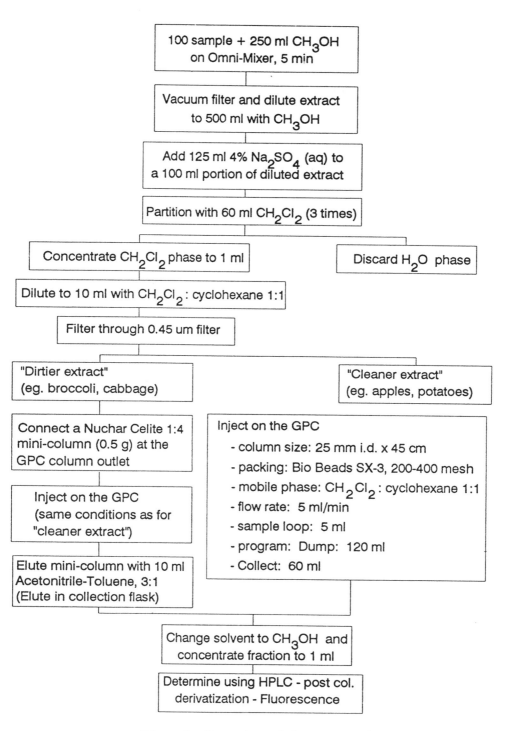

```
                ┌──────────────────────────────┐
                │  100 sample + 250 ml CH₃OH   │
                │     on Omni-Mixer, 5 min     │
                └──────────────────────────────┘
                ┌──────────────────────────────┐
                │  Vacuum filter and dilute extract │
                │     to 500 ml with CH₃OH     │
                └──────────────────────────────┘
                ┌──────────────────────────────┐
                │  Add 125 ml 4% Na₂SO₄ (aq) to │
                │  a 100 ml portion of diluted extract │
                └──────────────────────────────┘
                ┌──────────────────────────────┐
                │ Partition with 60 ml CH₂Cl₂ (3 times) │
                └──────────────────────────────┘
```

100 sample + 250 ml CH_3OH on Omni-Mixer, 5 min

Vacuum filter and dilute extract to 500 ml with CH_3OH

Add 125 ml 4% Na_2SO_4 (aq) to a 100 ml portion of diluted extract

Partition with 60 ml CH_2Cl_2 (3 times)

Concentrate CH_2Cl_2 phase to 1 ml

Discard H_2O phase

Dilute to 10 ml with CH_2Cl_2 : cyclohexane 1:1

Filter through 0.45 um filter

"Dirtier extract" (eg. broccoli, cabbage)

"Cleaner extract" (eg. apples, potatoes)

Connect a Nuchar Celite 1:4 mini-column (0.5 g) at the GPC column outlet

Inject on the GPC
  - column size: 25 mm i.d. x 45 cm
  - packing: Bio Beads SX-3, 200-400 mesh
  - mobile phase: CH_2Cl_2 : cyclohexane 1:1
  - flow rate: 5 ml/min
  - sample loop: 5 ml
  - program: Dump: 120 ml
  - Collect: 60 ml

Inject on the GPC (same conditions as for "cleaner extract")

Elute mini-column with 10 ml Acetonitrile-Toluene, 3:1 (Elute in collection flask)

Change solvent to CH_3OH and concentrate fraction to 1 ml

Determine using HPLC - post col. derivatization - Fluorescence

Scheme 1. Extraction and clean-up procedure.

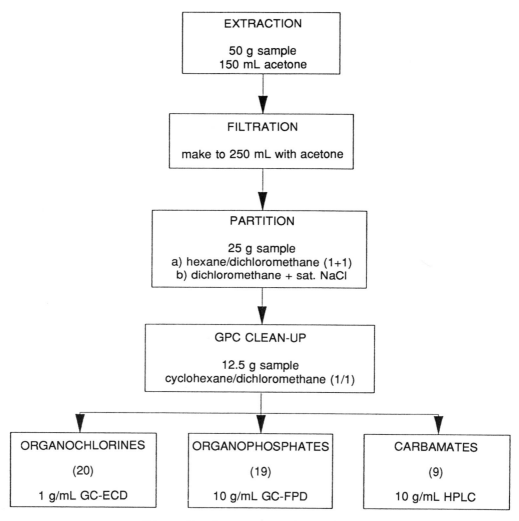

EXTRACTION

50 g sample
150 mL acetone

FILTRATION

make to 250 mL with acetone

PARTITION

25 g sample
a) hexane/dichloromethane (1+1)
b) dichloromethane + sat. NaCl

GPC CLEAN-UP

12.5 g sample
cyclohexane/dichloromethane (1/1)

ORGANOCHLORINES

(20)

1 g/mL GC-ECD

ORGANOPHOSPHATES

(19)

10 g/mL GC-FPD

CARBAMATES

(9)

10 g/mL HPLC

Scheme 2.   Acetone extraction method.

Table 1. Comparison of conventional detectors (CD) versus mass selective detector (MSD).

	CD 89-90	MSD 90-91
Analytical systems	3	2
Organochlorines	20	41
Organophosphates	19	49
N-containing (triazines)	-	25
Carbamates	9	10
Total actives	48	125
Mix standard solutions	5	1
Recoveries per set	20	125
On-line confirmation	No	Yes

Biobead SX-3 gel column and dichloromethane-cyclohexane (1:1 v/v) as eluting solvent. To further reduce analysis time, an Autovap attachment to the GPC automatically evaporated the column eluates at a rate of 5 ml/min.

This allowed the screening of approximately 50 pesticide residues or metabolites in a given sample matrix with limits of detection ranging from 0.01-0.08 ppm while recoveries varied from 72-110% depending on the compound. A further improvement in the multi-residue method for fruits, vegetables and soils was investigated, this involved extraction with acetone or acidic acetone in place of methanol (Scheme 2). Again, a partition step was involved together with GPC cleanup. The fractions from the GPC had to then be injected into different conventional detection systems, ie., OCs by GC-ECD, OPs by GC-FPD or GC-nitrogen phosphorus detecter (NPD) and the carbamates by HPLC. This in effect involved three analytical systems, quantification was lengthy and tedious, and if a positive result was obtained, further confirmation was needed, normally by mass spectrometry. Also, there was very limited quality assurance built into the procedure. In order to improve the quality and efficiency of analyses, the use of a mass selective detector (MSD) was adopted for samples requiring multi-residue screening. This procedure resulted in the use of two analytical systems, a simplified quantification process, QA data per residue per set and on-line confirmation. A comparison of the multi-residue method using conventional detectors versus MSD is shown in Table 1. In setting up the GC-MSD parameters, a target run time of one hour was chosen. The GC separation, typically on a 30 M x 0.25 mm DB-5 column using temperature programming, also had to be adequate to allow sufficient time for switching selective ion monitoring (SIM) groups throughout the run. The MSD gave a wide range of responses for the same amount of different residues, (by as much as a factor of 50) therefore, particular attention was given to the proper selection of target ions and their relative abundances. As can be seen, there were time savings as well as the scanning for 125 pesticides in one extraction. The applicability of the MSD to 76 pesticides analyzed by GC has been reported by Stan [5].

Recently, Anderson, from The Swedish National Food Administration, has advocated the use of ethyl acetate as extraction solvent in place of acetone. By using ethyl acetate it is possible to avoid the partitioning step which, in addition to being very time consuming, can also result in losses of the more water soluble pesticides (e.g., methamidophos, acephate, etc). Investigations are being carried out to compare the efficiency of these two different residue methods.

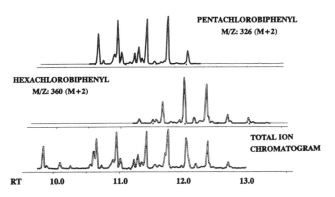

PENTACHLOROBIPHENYL
M/Z: 326 (M+2)

HEXACHLOROBIPHENYL
M/Z: 360 (M+2)

TOTAL ION
CHROMATOGRAM

RT 10.0 11.0 12.0 13.0

Figure 1.   Tomato extract spiked with Aroclar 1254 analyzed by GC/MS on a MAT-90 magnetic sector mass spectrometer at 2000 resolution. GC conditions: 70-270°C at 15°C/min on a 30 m DB-5 column, splitless injection.

PCB PROBLEM

In 1988, in Canada, there was a large PCB warehouse fire in St.-Basile-le-Grand, Québec. The resultant PCB cloud drifted for many days toward the St.-Lawrence River across agricultural land in which apple orchards, dairy farming, market gardening and corn production were possibly affected. A quick assessment of the situation was required due to the potential impact of PCBs on the residents of that community. Over 300 apples and other commodities were analyzed on a daily basis to assure their fitness for consumption. Initially, PCB residues were extracted with a mixture of acetone-hexane (1:1), cleaned-up on a florisil sep-pak with quantification on a temperature programmed 30 M DB-17 column using an ECD. Limits of detection varied from 1 to 5 ppb depending on substrate. Quantification was by summation of individual isomer peaks. Under the GC conditions used, all peaks for a 5 Arochlor mixture eluted between 5 and 17 minutes.

Therefore, all peaks on sample chromatograms were summed in this window. Like Durell and Sauer [7], results indicated a substantial number of peaks may be incorrectly identified in environmental samples with a one-column, one-detector system. However, for MS quantification, additional purification was necessary, resulting in a switch to GPC clean-up as well as SIM confirmation of each individual PCB isomer (Figure 1). The displayed ions (m/z 326 and 360) represent the most intense ions in the chlorine isotope cluster of the molecular ion of the respective chlorinated biphenyl congeners and show the ability to re-solve co-eluting compounds by mass. PCB residues were found as deposits on the skin of apples (range 1 to 39 ppb), on fresh corn (1 to 23 ppb), and on shell eggs (7 to 50 ppb). No residues were found in canned or processed vegetables.

The use of GC-MS or HPLC-MS (in the case of carbamates) has, in effect, shortened the time spent on various manual clean-up procedures or facilitated the acceptance of auto-mated sample preparation techniques such as GPC [8,9].

WATER ANALYSIS

Due to the use of agricultural pesticides, public concern has been raised in Canada concerning the possibility of drinking water and farm wells being contaminated by chemical leaching. To avoid the transportation of large volumes of water from the field sampling sites to the laboratory for low part per billion level analysis, an investigation was made into the use of a solid phase extraction (SPE) procedure. The objective was to develop a field sam-pling method, initially for triazine herbicides, that would meet the following guidelines:

Table 2. Determination of pesticide residues in water using the XAD-7 solid phase extraction procedure[a].

Pesticide	Recovery Levels (ppb)	Recovery Range (%)	Mean (%)	L.O.D.[b] (ppb)
Atrazine	0.5, 1, 10	91-120	102	0.1
Cyanazine	0.5, 1, 10	81-104	95	0.1
Metribuzin	0.5, 1, 10	80-119	94.6	0.1
Simazine	0.5, 1, 10	78-105	94.3	0.1
Alachlor	0.5, 1, 10	87-108	96.9	0.2
Metolachlor	0.5, 1, 10	85-104	95.2	0.2
Melathion	0.5, 1, 5	77-130	97	0.1
Azinphos-methyl	0.5, 1, 5	79-97	87.6	0.1
Dimethoate	0.5, 1, 5	69-105	88.9	0.1
Chlorpyrifos	0.5, 1, 5	77-109	90.6	0.1
Fenitrothion	0.5, 1, 5	80-118	94.1	0.1
Fonofos	0.5, 1, 5	70-106	89	0.1
Phorate sulphone	0.5, 1, 5	84-127	100.8	0.1
Trifluralin	0.5, 1, 2	44-81	65.1	0.2
Triallate	0.5, 1, 2	83-119	94.8	0.2

a) Minimise pesticide residues degradation without having to add preservative;
b) Easier storage and transportation than bottles;
c) Increase efficiency in the laboratory since a large number of samples were expected.

The criteria for the choice of resin was that the adsorption of pesticide had to be effective over a wide range of polarity, the resin should be coarse to minimize pressure drop in the cartridge, and have a large capacity. The adsorbents investigated were: Amberlite XAD-2, XAD-4, XAD-7 and C-18. XAD-2 and XAD-4 are styrene-divinyl benzene copolymers and function as non-polar hydrophobic adsorbents. Amberlite XAD-7 is an acrylic ester polymer, which is more polar and has somewhat more hydrophillic structure than XAD-2 and XAD-4 resins. XAD-7 gave the best recoveries for the compounds of interest which were of medium polarity. The C-18 resin was found to give back pressure problems in the cartridge due to the small bead size of the material, ie., 40 μM. Fifty 1.9 cm i.d. x 12 cm length cartridges were constructed of inert, unbreakable teflon for easy connection to water taps or garden hoses. The system was tested up to 2 litre throughput capacity and a sampling flow rate up to 125 ml/min with no adverse effect. At the field site, the volume throughput was recorded, the cartridge sent to the laboratory where the herbicide residues were eluted using 70 ml ethyl acetate and analysed by the normal multi-residue procedure and the cartridge returned for re-use. The range of pesticides, their recoveries and limits of detection are shown in Table 2. On average, a cartridge could be used for 15 samples, after regeneration with acetone and water, before being repacked. However, this depended upon the cleanliness of the water sample. In the initial study, 117 triazine and acetanilide residue samples were analysed and 14 were found to be positives with residue levels in the range of 0.2 to 10 ppb for atrazine, cyanazine, metolachlor and metribuzin. In a second-year study, 47 OP samples and 25 triallate and trifluralin samples were analysed by this procedure. Only 1 positive OP residue, namely dimethoate, was found at a level of 0.4 ppb and 2 positive samples for triallate and trifluralin residues at 0.2 and 0.3 ppb, respectively.

Attempts to use the XAD-7 cartridge with the phenolic/acidic pesticides, such as dinoseb, 2,4-D, etc. were unsuccessful. A 96 sample investigation for dinoseb in well water had to be carried out by the usual liquid-liquid water extraction procedure. Here, the water samples (400 ml) were acidified, extracted with dichloromethane followed by derivatisation with diazomethane and quantification by GC-ECD. In this study 19 positives were found for dinoseb in the range of 0.4 to 12.4 ppb. In summary, for water analysis, SPE columns have become the method of choice for sample extraction [10], as well as, clean-up [11]. Although the XAD-7 resin was unsuccessful with the more polar phenolic/acidic pesticides, Di Corcia [12] has reported a selective SPE procedure for 9 phenoxy acid herbicide residues in water using a mixed graphitized carbon black/silica-based anion exchange resin cartridge which gave >95% recoveries with limits of detection well below 0.1 µg/l (ppb).

NEW TECHNIQUES

With the objective of expanding the detection capabilities of multi-residue protocols, recent interest has been in the:

a) use of newer technologies, such as super-critical fluids for extraction and chromatography.
b) development of sample work-up procedures which can be undertaken by robotic systems. This results in greater throughput, at less cost, for high-volume sample programs.
c) advances in biotechnology have resulted in a limited number of "quick-test" screening procedures.

These areas present interesting possibilities for the improvement in time efficiency of a multi-residue pesticide method with ever lower detection limits. There has been a rapid growth in the use of supercritical fluid chromatography (SFE) during the 1980's [13,14]. The number of publications on the subject have gone from five in 1980 to 150 in 1989. The use of SFC and supercritical extraction (SFG) has been applied to soils, fruits and vegetables using a variety of detector systems, and it is claimed that 80% of the compounds that run on normal-phase liquid chromatography can be determined by SFC. We have investigated the determination of fluazifop in onions (Figure 2) by the conventional liquid extraction procedure using acidic acetone, filtration, derivatisation, ClinElut TM column clean-up using chloroform and GC-MS analysis. In total this procedure takes four hours analysis time. On-line SFE-SFC was attempted with both the acid and ester form of fluazifop using a number of different columns (ODS, cyano, PEG), but only the ester was observed. In comparison, Figure 3, shows the off-line SFE for the determination of ester and acid forms by HPLC-UV analysis resulting in a total analysis time of only 1.5 hours. Working in the range of 1 to 5 ppm ester, recoveries were greater than 70% without optimisation of all parameters. Further investigations included the use of methanol as modifier in the extraction vessal to improve the recovery of fluazifop acid to the same level as the ester. In addition, the application of SFC to the analysis of Linuron, Ethalfluralin and the triazine herbicide, Metribuzin and its metabolites in lupins is currently being assessed.

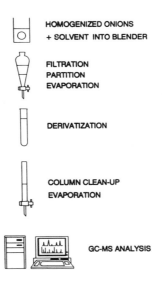

Figure 2.   Schematic for the analysis of Fluazifop by the conventional liquid extraction procedure.

RAPID DIAGNOSTIC TESTS

In recent years, immunoassay technology has found applications in environmental pesticide residue testing primarily in water and soil substrates [15,16]. The method is based upon the specific binding nature of monoclonal and polyclonal antibodies. These proteins can be custom generated such that they will bind only with the specific pesticide molecules of concern. The technique is rapid, inexpensive, and requires little personnel training. However, multi-residue determinations are not possible unless the target compounds are structurally related, also these immunoassay test kits do not have the ability to distinguish between these structurally related chemicals. In the last two years, we have investigated a

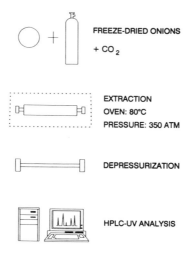

Figure 3.   Schematic for supercritical fluid extraction of Fluazifop.

Table 3. Application of the rapid immunoassay test kit to plant matrices.

Compound	Sensitivity (PPB) water & soil	Sensitivity (PPB) fruit & vegetable	Sensitivity loss
Endosulfan	8	50	x 6
Endrin	7	50	x 7
Dieldrin	20	100	x 5
Heptachlor	45	1000	x 22

number of systems as possible rapid field tests for use by agriculture inspectors in remote locations. These have included the Enzytec System, Ez-Screen Test System, the Res-I-Mune TM System, and a Pentachlorophenol Field Analysis Kit. A previous report [17] has been made on the applicability of pentachlorophenol to water and animal tissue samples as well as the raising of polyclonal antibodies to fluazifop in the development of an ELISA screening procedure. More recently, the advantages and limitations of the Res-I-Mune TM organochlorine test kit were compared with those of the multi-residue method for fruits and vegetables described above. Sample preparation involved the 250 ml acetone extraction of a 50 g sample, evaporation of 125 ml extract (25 g sample) followed by dilution with 25 ml sodium phosphate buffer/tween solution. Colour development, carried out using the manufacturer's instructions, was accomplished in 7 minutes with absorbance readings on the resulting yellow solution being measured at 450 nm using a UV/visible spectrometer. Alternately, a visual comparison could be performed using a range of spiked controls. In all, the complete assay, from crop preparation to detector, took less than 20 minutes. The organochlorine residues tested were all structurally related, being of the cyclodiene type. The limit of detection (LOD) was defined as three times the standard deviation at the lowest fortification level. The LODs were 0.02 ppm for endosulfan in tomato, lettuce and apple and for endrin in apple and 0.05 ppm for dieldrin in apple using the criteria of 24 to 27% inhibition. A matrix effect for endosulfan in the three different crops was not observed. Cross reactivity of heptachlor was observed at 1 ppm showing a 50% inhibition (I_{50}) from the control sample. A comparison of assay sensitivity at I_{50} between substrates is shown in Table 3. The transition from water and soil to fruit and vegetable samples resulted in a loss in sensitivity, ranging from 6 to 22 times less sensitive.

In conclusion, the kit was found to be simple, rapid (each assay took less than 20 minutes) and inexpensive to detect endosulfan, endrin and dieldrin in fruits and vegetables with LODs of 0.02 to 0.05 ppm and with no false negative results.

ACKNOWLEDGMENT

The authors would like to thank M. Lanouette, R. Grant, G. Malis and R. Hindle for their assistance.

REFERENCES

1. P.A. Mills, J.H. Onley and R.A. Gailher. Rapid Method for Chlorinated Pesticide Residues in Non-fatty Foods. *J. Assoc. Off. Anal. Chem.* 46, 186-191 (1963).
2. Pesticide Analytical Manual: U.S. Food and Drug Administration, Rockville, MD, Vol. 1 (Sections 211 and 212). June 1990.

3.   R.T. Krause. Multi-residue Method for Determination of N-Methyl Carbamate insecticides in Crops, using High Performance Liquid Chromatography. *J. Assoc. Off. Anal. Chem.*, 63, 1114-1124 (1980).

4.   D. Chaput. Simplified Multi-Residue Method for Liquid Chromatographic Determination of N-Methyl Carbamate Insecticides in Fruits and Vegetables. *J. Assoc. Off. Anal. Chem.*, 71, 542-546 (1988).

5.   H.-J. Stan. Application of Capillary Gas Chromatography with Mass Selective Detection to Pesticide Residue Analysis. *J. Chromatography*, 467, 85-98 (1989).

6.   A. Anderson, H. Pälsheden. A Capillary Gas Chromatographic Multi-residue Method with Ethyl Acetate Extraction and GPC Clean-up for the Determination of Pesticides in Fruits and Vegetables. *Fresenius J. Anal. Chem.*, 399, 365-367 (1991).

7.   G.S. Durell and T.C. Sauer. Simultaneous Dual-Column, Dual-Detector Gas Chromatographic Determination of Chlorinated Pesticides and Polychlorinated Biphenyls in Environmental Samples. *Anal. Chem.*, 62, 1867-1871 (1990).

8.   J.M. Czuczwa, A. Alfond-Stevens. Optimised GPC Clean-up for Soil, Sediment, Water and Oily Waste Extracts for Determination of Semivolatile Organic Pollutants and PCBs. *J. Assoc. Off. Anal. Chem.*, 72, 752-759 (1989).

9.   S.J. Chamberlain, Determination of Multi-Pesticide Residues in Cereals, Cereal Products and Animal Feed using Gel-permeation Chromatography. *Analyst (London)*, 115, 1161-1165 (1990).

10.   M.W. Brooks, J. Jenkins, M. Jimenez, T. Quinn and J.M Clark. Rapid Method for the Determination of Alachlor, Atrazine and Metalachlor in Ground Water by Solid-phase Extraction. *Analyst (London)*, 114, 405-406 (1989).

11.   W.-K. Wang, S.-D. Huang. Rapid Determination of Seven Herbicides in Water or Iso-Octane using C_{18} and Florisil Sep-Pak Cartridges and Gas Chromatography with Electron-Capture Detection. *J. Chromatogr.* 483, 121-129 (1989).

12.   A. Di Corcia, M. Marchette and R. Samperi. Extraction and Isolation of Phenoxy Acid Herbicides in Environmental Waters using Two Absorbents in one Mini Cartridge. *Anal. Chem.*, 61, 1363-1367 (1989).

13.   M.L. Lee and K.E. Markides. "Analytical Supercritical Fluid Chromatography and Extraction", Chromatography Conferences Inc., Provo, Utah, USA (1990).

14.   D.E. Knowles, B.E. Richter, M.R. Anderson and D.W. Later, Supercritical Fluid Chromatography: A New Approach to Pesticide Analysis. *Chemi. Oggi*, 7, 11-15 (1989).

15.   R.J. Bushway, B. Perkins, S.A. Savage, S.J. Lekousi and B.S. Ferguson. Determination of Atrazine Residues in Water and Soil by Enzyme Immunoassay. *Bull. Environ. Contam. Toxicol.*, 40, 647-654 (1988).

16.   E.M. Thurman, M. Meyer, M. Pues, C.A. Perry and A.P. Schwab. Enzyme-Linked Immunosorbent Assay compared with GC/MS for the Determination of Triazine Herbicides in Water. *Anal. Chem.*, 62, 2043-2048 (1990).

17.   W.P. Cochrane. Testing of Food and Agricultural Products by Immunoassay - Recent Advances, in: "Immunoassays for Trace Chemical Analysis". M. Vaderlaan, L.H. Stanker, B.E. Watkins and D.W. Roberts eds. ACS Symposium Series 451, American Chemical Society, Washington, D.C., USA (1990).

A NOTE ON SUPERCRITICAL FLUID EXTRACTION (SFE) - AN ALTERNATIVE TO TRADITIONAL TECHNIQUES IN REGULATORY ENVIRONMENTAL ANALYSES?

A.J. Penwell and M.H.I. Comber

ICI Group Environmental Laboratory
Freshwater Quarry
Brixham, Devon
TQ5 8BA, UK.

INTRODUCTION

The extraction and recovery of organic molecules from environmental matrices such as tissues and sediments is a critical step in the investigation of the environmental behaviour of such molecules.

Regulatory environmental analysis as discussed here covers two broad areas:-
a) The determination of single components, frequently pesticides, in a variety of matrices (often including fish).
b) Investigation into the behaviour of organics in the environment.

Current accepted methods for the extraction of polar molecules from environmental samples involve liquid extractions by Soxhlet or sonication techniques. Such extractions are lengthy, commonly taking a number of hours to perform. Relatively large volumes of ultra-pure solvent are needed which is both expensive and undesirable for reasons of safety. Furthermore, the extractions sometimes yield incomplete recovery, often need subsequent concentration steps and may lead to degradation of the target species.

The object of this investigation was to assess the claimed advantages of supercritical fluid extraction (SFE). Briefly, these are that SFE yields rapid, reproducible and selective extraction whilst being convenient and easy to automate. Extracts can be easily analysed both on and off-line. Solvent handling is minimal and the fluid may be simply separated from the extract by decompression, allowing for preconcentration of the extract for trace analysis (1,2). It was hoped that the technique would yield these improvements within a regulatory environmental laboratory.

EXPERIMENTAL

The supercritical fluid extractions were performed on a modified Gilson system, or a Jasco system marketed by Ciba-Corning. The modified Gilson system comprised the follow-

Sample Preparation for Biomedical and Environmental Analysis,
Edited by D. Stevenson and I.D. Wilson, Plenum Press, New York, 1994

203

ing units: A Techne circulating refrigerated bath; a heat exchanger on the pump inlet line; a Gilson 305 modifier pump; a Gilson 303 carbon dioxide pump; a Gilson 811 dynamic mixer; a Gilson 802Ti manometric module; an ICI Tc1900 HPLC oven; two rheodyne 6 way valves; a Rheodyne 7037 pressure relief valve and Capital 10 cm HPLC cartridges (Capital HPLC specialists). The Ciba-Corning system was used as purchased and comprised of modified Jasco LC pumps for modifier and carbon dioxide; a Jasco HPLC oven; two 3-position switching valves; a Jasco 875 UV detector and a Jasco back pressure regulator (BPR).

For extraction of the pesticide on the Gilson system the conditions were: 5% ethanol in carbon dioxide, with a flowrate of 4 ml/min, and oven temperature of 45°C. The back pressure was set at 2000 psi. Two 5 minute extraction periods were run consecutively and the extracts analysed separately. For the Jasco system the conditions were similar except that the BPR was set at 138 kg/cm^2 and 30°C. Further extracts were run on the Jasco system with the following conditions: 5% ethanol in carbon dioxide, with a flow rate of 1.0 ml/min and an oven temperature of 60°C. The BPR was operated at 148 kg/cm^2 and 60°C. Two dynamic extraction periods of 15 minutes, run consecutively, were performed using these conditions. Extractions of surfactants were made on the Jasco system under the following conditions: carbon dioxide only, or with 5% ethanol, at a flow rate of 2.0 ml/min and an oven temperature of 60°C with the BPR set at 138 kg/cm^2 at 60°C. Two dynamic extraction periods each of 15 minutes were performed under these conditions.

Extraction Of Pesticide From Fish

The pesticide used was an alkanoic acid based post emergence herbicide. The octanol-water partition coefficient is greater than 3 which suggests that the herbicide has the potential to accumulate in fish tissue. The compound has an extremely low pKa and is therefore unlikely to dissociate in the environment.

The Mirror carp *(Cyprinus carpio)* was chosen for the investigation of the extraction of pesticide from fish tissue as this is frequently the preferred species in bioaccumulation tests, e.g. MITI, Japan. Fish of between 7 and 10 g were cleanly killed, weighed and spiked at 100mg/kg by injection of an appropriate volume of a solution of the pesticide in acetone into the flank and gut by means of a syringe. Thorough maceration of the fish was achieved by low and high speed chopping with 40ml acetone in a Waring blender. The fish tissue was transferred to a weighed petri dish and the weight of the fish tissue recorded. The uncovered dishes were stored overnight until constant weight had been achieved, generally 20% of the original weight. The dried material was stored in a freezer below -20°C until required for extraction as described above. All extracts were blown to dryness under a stream of nitrogen and redissolved by sonication in 5ml hexane for analysis.

The liquid chromatographic analysis of the pesticide extracts was carried out on a system comprising a Waters 510 LC pump, Perkin Elmer ISS-100 Autosampler and Uvikon 740UV detector. Data collection and manipulation was performed using Waters Maxima 820 software. The chromatographic conditions were: 12.5cm x 4.9mm i.d. column packed with covalent bonded D-Phenylglycine, mobile phase 3% ethanol in hexane at 2.5ml/min, with UV detection at 254nm. The injection volume was 200µl.

Extraction Of Pesticide From Sediment

Pesticide spiked sediments were prepared at 500mg/kg by the addition of an appropriate volume of pesticide in acetonitrile to a known weight of standard sediment which was mixed well and left overnight to dry. Before use, the sediment was ground thoroughly in a pestle and mortar to ensure even distribution of the pesticide. Preparation and analysis of the extracts was performed as described above.

Extraction Of Surfactants From Fish

Surfactant spiked Mirror carp were prepared at 200mg/kg by direct injection of a known volume of the surfactants Nonyl phenol ethoxylate (NPE) and C-12/C-14 alcohol ethoxylates in 50:50 methanol:water into a fish of known weight. The tissue was macerated as described previously.

The extracts were acidified and blown to dryness at 100°C under a stream of nitrogen. Internal standard (C-8/C-20) and dichloroethane and phenyl isocyanate derivatisation agents were added to the residue. After derivatisation at 70°C for 16 hours the mixture was again dried and taken up in 500μl chloroform for analysis.

Table 1.   Recovery of a pesticide from fish tissue.

1st or 2nd extraction	Weight extracted	% recovery	% total recovery
1*	0.611 g	18	41
2*		23	
1	0.428 g	31	38
2		7	
1	0.381 g	33	38
2		5	
1	0.412 g	46	49
2		3	
1	0.479 g	46	47
2		1	
1	0.454 g	48	51
2		3	
1	0.501 g	47	53
2		6	
1	0.495 g	51	54
2		3	

* denotes data obtained on Gilson system (wet sample).
All other data for extractions from Jasco system.

The chromatographic analysis of the surfactant extracts was carried out on a system comprising two Waters 510 LC pumps, a Perkin Elmer ISS-200 Autosampler, a Jasco 735 UV detector and a Waters Automated gradient controller. Data manipulation was performed using Maxima 820 software.

The chromatographic conditions were: Column 15cm x 2.1mm i.d. Waters Bondapak C18, mobile phase 70:30 methanol:water for 1 minute increased to 100% methanol over 20 minutes at 1ml/min, with UV detection at 240nm. The injection volume was 5μl.

Table 2.   Recovery of a pesticide from sediment.

1st or 2nd extraction	Weight extracted	% recovery	% total recovery
1	1.623	40	43
2		3	
1	0.454 g	9	9.4
2		0.4	
1	0.436 g	56	56

Although a thorough investigation of extraction conditions was not made, it is clear that the second extraction period removed little more of the target compound than the first five minutes extraction. This is explained by considering the relationship between solubility and pressure (3) and does suggest consistency of a single extract at a fixed pressure. The method could therefore have application in the regulatory environment recovery. Whilst the overall percentage recovered does not approach 100%, this is often not the case for liquid extractions either. Higher pressures would be necessary for complete extraction.

SFE held clear advantages over traditional solvent methods for although the preparation of dried samples was relatively lengthy, overnight drying was successful. Once prepared, 18 samples were extracted in five and a half hours. At only 18 minutes per sample, this would represent a considerable overall advantage compared with liquid extraction. Also, the conditions of SFE are much less aggressive than liquid extraction and might therefore be beneficial for thermally labile compounds.

The cleanliness of control samples suggests that a considerable reduction in background interference by selectivity of the extraction would be a real possibility in the regulatory analysis of fish tissue by SFE. Concentration steps for trace analysis are minimised for although levels of the analyte were relatively high, the extract could be collected into relatively small volumes of the analysis solvent.

The data presented in Table 2 give recoveries of the pesticide from spiked sediments. The percentage recoveries obtained are similar to the fish tissue samples discussed above.

The advantages of SFE over the liquid extraction methods for sediments would be greatest in terms of the selectivity which is possible by control of the pressure and hence density of the supercritical fluid (4). In terms of the regulatory studies this selectivity would

Table 3.   Recovery of surfactants from fish.

Extraction conditions	% recovery Nonyl Phenol Ethoxylate	% recovery C12/C14 Alcohol Ethoxylate
No EtOH, 2000 psi	4	14
No EtOH, 4000 psi	51	61
+ EtOH, 2000 psi	40	54
+ EtOH, 4000 psi	30	42

Note: EtOH = Ethanol

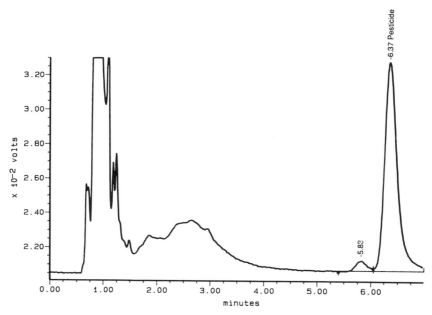

Figure 1. HPLC chromatogram of SFE extract of pesticide from fish, conditions in text.

result in cleaner extracts reducing the problems of background interferences commonly experienced. Furthermore the use of SFE would reduce the need for large volumes of organic solvents thereby reducing costs and safety considerations. A further advantage is that while soxhlet extraction often uses 50g samples, SFE was performed with <0.5g thus multiple extractions to assess the reproducibility of the technique were easier. This would be particularly valuable where the sample size was limited, for example in this study carp of between 7 and 10 g were used.

The data in Table 3 illustrates the potential selectivity of supercritical fluids as extraction media. Using pure CO_2 only, an increase in pressure and hence increased density of the fluid is reflected by a marked increase in the recovery of both surfactants. The addition of ethanol to the CO_2 dramatically increased recovery at 2000 psi due to the increased polarity of the fluid, whilst increasing the pressure recovered more of target compounds. However, in developing a method of this kind, care must also be taken not to increase the pressure to levels at which unwanted materials such as lipids are coextracted. In this experiment a small but significant increase in the late eluting non-surfactant peaks was observed when the pressure was increased. The total recoveries were good, and the selectivity of increased polarity can be seen from Figures 2 and 3, as the low density low polarity conditions of Figure 2 extracted a different range of compounds, whilst obviously the subsequent more polar conditions with modifier recovered more of the target surfactants.

Further optimisation of these extractions would need to investigate more fully the effects of pressure and hence fluid density. Initial results of other workers (4) have demonstrated density programming may yield better results. The initial preparation and physical state of the sample would be investigated as there is clear evidence that the state of the sample and other modifiers within it may influence the extraction (5).

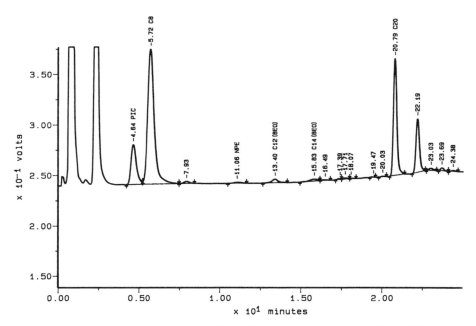

Figure 2. HPLC chromatogram of SFE of surfactant from fish, no modifier 2000 psi.

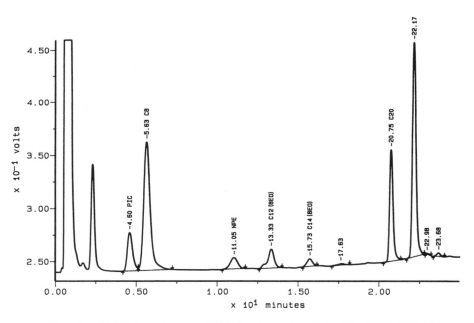

Figure 3. HPLC chromatogram of SFE of surfactant from fish, with ethanol, 2000 psi.

REFERENCES

1. S.B. Hawthorn. Analytical-scale supercritical fluid extraction. *Anal. Chem.*, 62, 633A-642A (1990).
2. M. Lohleit, R. Hillman and K. Bachman. Comparison of the efficiency of different GLC multi-residue methods on crops containing pesticide residues. *Fres. J. Anal. Chem.*, 339, 470-474 (1991).
3. K.D. Bartle and A.A. Clifford. Can supercritical-fluid extraction live up to its promise? *LCGC Interna tional.*, 10, 10-12.
4. M. Kane, J.R. Dean, S.M. Hitchen, C.J. Dowle and R.L. Tranter. Supercritical fluid extraction as a sample preparation technique for chromatography and spectroscopy. *Anal. Proc.*, 29, 31-33 (1992).
5. M.L. Hopper and J.W. King. U.S. Patent application PAT-APPL-7-536861. Available from the National Technical Information Service, Springfield, Virginia 22161, U.S.A.

A NOTE ON SAMPLING AND ANALYSIS OF VOLATILE ORGANICS USING

AUTOMATED THERMAL DESORPTION

R.J.Briggs and D.Stevenson

Robens Institute of Health and Safety
University of Surrey
Guildford,
Surrey
GU2 5XH. UK

INTRODUCTION

The introduction of various pieces of legislation concerned with health and safety at work, along with increasing public concern about possible adverse health effects caused by human exposure to low levels of toxic chemicals, has increased the need to carry out work-place monitoring. Knowledge of the risk to human health of handling toxic chemicals clear-ly needs measurement of the exact levels that humans are exposed to. The vast number of chemicals now encountered in the workplace means that no single sampling strategy and no single analytical technique is applicable to all situations. Although the development of direct reading instruments for on-site monitoring, giving fairly instant results, is progressing the majority of measurements of trace organic compounds in workplace samples require collect-ing samples in the workplace and sending to an analytical laboratory for instrumental analy-sis.

The inherent specificity and sensitivity of chromatographic methods means that they are extensively used for workplace monitoring of organic compounds. Gas liquid (both packed column and capillary), and high performance liquid chromatography (HPLC) includ-ing ion chromatography, are extensively used and thin layer chromatography (TLC) can occasionally offer an alternative. The major limitation of gas chromatography (GC) for many applications, (the requirement for volatility), is much less of a problem for workplace monitoring, where the sample matrix is often air and the analyte is a volatile organic com-pound. This technique is thus extensively used.

SAMPLING

A range of sampling methods have been used. The exact strategy depends on both the chemical nature of the compound(s) to be sampled and the work operation that has to be monitored, which in itself can affect the matrix which subsequently has to be analysed.

Sample Preparation for Biomedical and Environmental Analysis,
Edited by D. Stevenson and I.D. Wilson, Plenum Press, New York, 1994

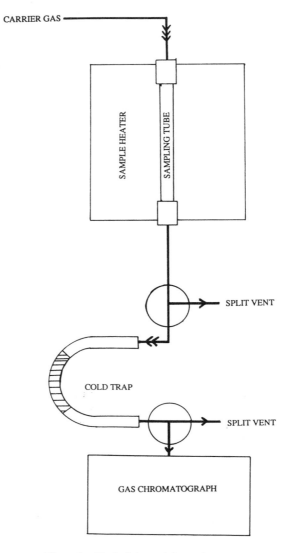

CARRIER GAS

SAMPLE HEATER

SAMPLING TUBE

SPLIT VENT

COLD TRAP

SPLIT VENT

GAS CHROMATOGRAPH

Figure 1.   Typical thermal desorption apparatus.

One of the approaches most extensively used for sampling organic vapours in the workplace involves pulling air through a tube containing a solid adsorbent placed in the breathing zone of a worker. Air is sampled at a known flow rate for a set time period (often the daily workshift). The compound(s) of interest are trapped on the surface of the sorbent and then later analysed by GC after desorption from the sorbent. The most commonly used sorbent is charcoal though silica gel, alumina and a variety of porous polymers are also used. Carbon disulphide is most often used as the desorbing solvent with charcoal [1]. It is an effective solvent for recovering analytes from the charcoal and gives quite a low response to the flame ionisation detector used with GC. Unfortunately it is a fairly toxic material and needs careful handling. A safer and more direct approach to desorption is to perform thermal desorption of the sorbent, thus eliminating the need for the use of hazardous solvents. Adsorbent tubes are cheap, convenient to use and this approach in combination with GC can give detection limits in the parts per million range.

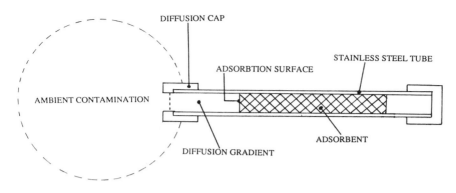

Figure 2.   Thermal desorption tube.

PRINCIPLES OF THERMAL DESORPTION

The basic principles of thermal desorption are relatively simple. The matrix containing the analytes of interest is heated to a temperature which allows these volatile components to be given off as a vapour. This vapour is then subjected to separation into its individual components using a suitable GC column and detected in the normal way. Two major problems present themselves.

Firstly ensuring that all, or a reproducible percentage of the components is freed from the matrix. This often involves long periods of heating at elevated temperatures, particularly in the case of solid or semi solid matrices such as tablets or biological tissues, which often have to be crushed or macerated prior to analysis to facilitate this process.

The second problem is allied to the first in that as the analyte is released over an extended period of time, in order for any meaningful chromatography to be conducted the sample must be condensed and presented for analysis as a more discrete sample. This is achieved either by cryogenic focusing at the head of the column, or by trapping the sample on an adsorbent of a much smaller size. In either case the sample is flash volatilised into the GC, thereby reducing peak spreading to a minimum.

The method of adsorbing onto a particulate adsorbent bed is in many ways preferable to cryogenic focusing on column because higher water vapour contents in the sample may cause blocking of the capillary column in the later procedure, leading to imprecision in injection.

A diagram of a typical thermal desorption apparatus is given in Figure 1. This illustrates the basic design of the Perkin Elmer ATD-50, an automatic multisample thermal desorption unit. This system is equipped with splitments both before and after the cold trap, thus allowing for optimal concentrations of analyte to be held on the cold trap and injected onto the GC. A typical thermal desorption tube is shown in Figure 2.

APPLICATIONS OF THERMAL DESORPTION

Gaseous Matrices

This is the most common application for thermal desorption analysis and is extensively used in the fields of occupational hygiene and environmental assessment to measure traces of potentially harmful organic contaminants in workplaces or in emissions into the environment [2,3,4,5]. The attraction of this method of sampling compared with other accepted techniques such as chemical desorbtion from adsorption tubes (charcoal or silica gel) is that the samplers are amenable to direct analysis without any further handling. This increases the speed of analysis, reduces costs and reduces the number of possible sources of error.

Sampling can be conducted in two modes, either pumped or diffusive. In both cases the sampling tube must be packed with an adsorbent material that will ensure that the compounds of interest are trapped efficiently but also allow them to be released at temperatures that will not degrade the sample.

Pumped sampling requires a known volume of the atmosphere to be sampled to be drawn through the adsorbent material. The advantage of this method is that when using precise, calibrated sampling pumps extremely accurate determinations of very low atmospheric concentrations of organic compounds can be performed.

Diffusive sampling, on the other hand, relies on the principle that the diffusion gradient from the surface of the adsorbent to the atmosphere to be assayed allows a constant sampling to be achieved. In theory the uptake rate of the compounds of interest should be constant and related to the diffusion coefficient by Fick's law. However this requires a perfect adsorbent, namely one that totally adsorbs giving a zero concentration at the surface. In practice, although many adsorbents have been developed which approximate to this for particular analytes, no adsorbent is totally ideal. In addition uptake rates may be affected by such physical parameters as air velocity across the face of the tube, temperature, humidity, etc. These problems necessitate extensive checking and validation of every proposed analyte - adsorbent sampling system to ensure accurate determinations. The advantages offered by this type of sampling, in particular the ease of sampling when considering multiple site monitoring or monitoring the exposure of individual workers to hazardous materials, often makes the extra method development costs worthwhile [6].

The techniques described in the preceding paragraph are not restricted to the analysis of air but have been used to measure contaminants and components in other gas samples, such as natural gas, and samples from landfill sites [7].

Liquid Matrices

A number of techniques involving thermal desorption and GC as an end step can be applied to the determination of organic components in liquid matrices. The simplest technique is direct thermal desorption of an aliquot of the sample. The aliquot is spiked directly onto a solid support, either glass wool or a porous polymer packing material. This is normally placed within a PTFE liner in the stainless steel thermal desorption tube to ensure that there is no decomposition of the components on the metal surface. While this technique is applicable to a range of liquids, water based samples have problems due to the large volume of water which may be taken onto the column. In these cases it is sometimes possible to identify a cold trap packing which only adsorbs the compounds of interest while allowing water to be purged. Final desorption could then be performed without risk of overloading the column with water [8,9].

Flavour and fragrance analysis of liquids can be performed by headspace analysis by drawing the headspace onto a suitable adsorbent packing prior to thermal desorption analysis. This method allows all the available sample to be trapped for analysis and therefore aids sensitivity. One disadvantage of this type of analysis is that results may be highly matrix dependent, with a different extraction profile from sample to sample. The method is therefore appropriate for quality control work in beers, wines, spirits and other materials which are very similar from sample to sample [10,11], but may be prone to error in samples such as river waters, sediments and similar types of samples, where the matrix may vary considerably.

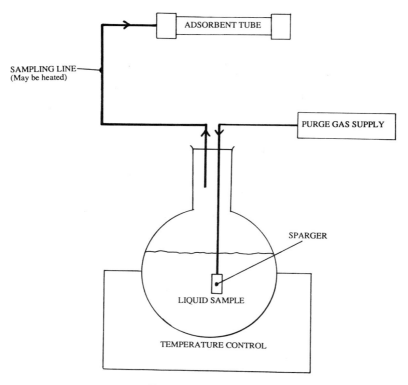

ADSORBENT TUBE

SAMPLING LINE—
(May be heated)

PURGE GAS SUPPLY

SPARGER

LIQUID SAMPLE

TEMPERATURE CONTROL

Figure 3. Open loop system.

Where matrix differences do occur, or where additional sensitivity is required purge and trap techniques are more applicable. With these techniques the sample is actively sparged to encourage the volatalisation of the analytes, which are then swept onto a suitable adsorbent packing [12,13]. This trapping may be performed in open or closed loop systems each of which has its advantages and pitfalls. The open loop system (Figure 3) purges the sample using a gas supply which is then vented to exhaust following the adsorbent trap. The closed loop system relies on the gas being pumped around the system and continually cycling through the packing. The latter system, by using the same gas supply, ensures that contamination from purge gas is reduced to a minimum. However, problems may occur due to contamination from the pump system, or because of equilibra set up in the system which may lead to partition of the analytes between the adsorbent packing and the vapour saturated gas. The former system all but eliminates these sources of error but the purge gas must be extremely pure to ensure that significant contamination is not introduced.

Solid And Semi Solid Matrices

Thermal desorption has been used to extract and measure volatile organics in a variety of solid matrices and in essence the solid material is treated in a similar way to the adsorbent packing used in air sampling. The dried sample is placed within a stainless steel tube, heated and the vapours released are swept onto and trapped by a suitable cold trap.

This technique has been used to measure residual solvents in drugs, volatiles in packing materials [14], residual solvents in polymer resins [15,16,17], petrol in arson residues [18] and many other similar applications.

The suitability of the method relies to a great extent on the stability of the matrix. Obviously if this is thermally labile under conditions required for full extraction then the potential for overloading the cold trap or column with breakdown products exists.

Problems of calibration and validation of the methods are no more difficult than for solvent extraction methods and the thermal desorbtion approach is amenable to method development as repeated extraction of the same sample to determine recoveries requires only the re-desorbtion of the previously prepared tube.

CONCLUSION

Thermal desorption techniques are thus established as one of the best methods for handling air samples. Ease of automation means that the technique is attractive for sampling volatile organics in other sample types.

REFERENCES

1. Health and Safety Executive. Methods for the Determination of Hazardous Substances. Benzene in air. Laboratory method using charcoal adsorbent tubes, solvent desorption and gas chromatography. MDHS 17. HSE London, 1987.
2. Health and Safety Executive. Methods for the Determination of Hazardous Substances. Benzene in air. Laboratory method using pumped porous polymer adsorbent tubes, thermal desorbtion and gas chromatography. MDHS 22. HSE London, 1990.
3. Health and Safety Executive. Methods for the Determination of Hazardous Substances. Styrene in air. Laboratory method using porous polymer diffusive samples, thermal desorption and gas chromatography. MDHS 43. HSE London, 1985.
4. Health and Safety Executive. Methods for the Determination of Hazardous Substances. 1,3-butadiene in air. Laboratory method using molecular sieve diffusion samples, thermal desorption and gas chromatography. MDHS 63. HSE London, 1989.
5. P.T.Crisp, J.Ellis, J.W.De Leeuw, and P.A.Shenk. Flash thermal desorption as an alternative to solvent extraction for the determination of C8-C35 hydrocarbons in oil shales. *Anal. Chem.*, 58(1), 258-261 (1986).
6. Health and Safety Executive. Methods for the Determination of Hazardous Substances. Protocol for assessing the performance of a diffusive sampler. MDHS 27. HSE London, 1983.
7. O. Janson. Analysis of trace pollutants in landfill gas. *G.I.T. Fachz. Lab.*, 33(6), 558-568 (1989).
8. J.J.Vreuls, U.A.Th.Brinkman, G.T.De Jong, K.Grob and A.Antho. On-line solid phase extraction-thermal desorption for introduction of large volumes of aqueous samples into a gas chromatograph. *J. High Resolut. Chromatogr.*, 14(7), 455-459 (1991).
9. Z.Voznakova and M.Popl. Sorption of phenols from water and subsequent thermal desorption for gas chromatographic analysis. *J. Chromatogr. Sci.*, 17(12), 682-686 (1979).
10. S.M.Belledeau, B.J.Miller and H.C.Thompson. N-nitrosamine analysis in beer using thermal desorption injection coupled with GC-TEA. *J. Food Sci.*, 53(6), 1696-1698 (1988).
11. E.Benfati, M.Natangelo, E.Daushi and R.Fanelli. Migration of vinyl chloride into PVC bottled drinking water assessed by gas chromatography mass spectrometry. *Food and Chemical Toxicol.*, 29(2), 131-134 (1991).
12. A.Nishimura, K.Kondo, E.Nakazawa, H.Mishima and S.Takenura. Identification and measurement of 'mureka' in sake. *Hakkagaku Kaishi*, 67(4), 237-244 (1989).
13. I.Pons, C.Garrault, J.N.Jaubert, J.Morel and J.C.Fenyo. Analysis of aromatic caramel. *Food Chemistry*, 39(3), 311-320 (1991).
14. J.Gilbert, J.M.Ingram, M.P.Scott and M.Underhill. The analysis of clingfilms by infrared spectroscopy and thermal desorption capillary gas chromatography. *J. Forensic Sci. Soc.*, 31, 337-347 (1991).
15. A.Ya.Lazaris, S.M.Schmujlovich and T.A.Kamikova. Determination of residual vinyl chloride in PVC resins. *J. Chromatogr.*, 198(3), 337-346 (1980).
16. P.J.Taylor, D.Price, G.T.Milnes, J.M.Scrivens and T.G.Blease. Thermal desorption gas chromatography mass spectrometry studies of commercial polypropylene samples. *Int. J. Mass Spectrum. Ion. Processes*, 89(2-3), 157-169 (1989).
17. J.R.Myers. Gas chromatographic analysis of trapped solvents in polymers by direct thermal desorption. *Am. Lab.*, 14(12), 36-39 (1982).

18. M.Frenkel, S.Tscroom, Z.Aizenshtat, S.Kraus and D.Daphna. Enhances sensitivity in analysis of arson residues: an adsorption tube - gas chromatograph method. *J. Forensic Sci.*, 29(3), 723-731 (1984).

SOLID-PHASE EXTRACTION PROCEDURE FOR THE ASSAY OF 1,4-DIOXANE IN

COSMETIC PRODUCTS

S. Scalia and G. Frisina[1]

Department of Pharmaceutical Sciences
University of Ferrara
via Scandiana 21
44100 Ferrara
Italy

[1]Himont Research Centre
Ferrara
Italy

SUMMARY

Cosmetic preparations containing polyethoxylated surfactants can be contaminated by 1,4-dioxane. This substance has been shown to be carcinogenic to rats and mice and has been considered as a possible carcinogen in humans. According to the EEC directive on cosmetic products, 1,4-dioxane must not be present in their formulation. Published methods for the assay of this substance in cosmetics are based on gas chromatography (GC) or head-space GC. These techniques have distinct disadvantages including complex and time-consuming sample pre-treatment, extensive calibrations, poor accuracy and precision. Improved recovery and reproducibility have been attained by a recently developed HPLC method; however this technique exhibits a low sensitivity. This study describes a new solid-phase extraction procedure for the determination of 1,4-dioxane in cosmetic preparations. After purification by Bakerbond C18 and CN cartridges, samples were directly analysed by HPLC and GC-MS. The proposed method is rapid, reproducible and suitable for routine quality control analyses. The application of the procedure to the assay of 1,4-dioxane in a wide range of commercially available surfactants and cosmetic products is also reported.

INTRODUCTION

Polyethoxylated derivatives are the most widely used surfactants in shampoo and bath preparations [1] and are commonly contained in other cosmetic products [2]. 1,4-dioxane may be formed during the polymerisation of ethylene oxide to produce the polyoxyethylene moiety of the emulsifiers [3,4]. Hence, cosmetics containing ethoxylated surfact-

Sample Preparation for Biomedical and Environmental Analysis,
Edited by D. Stevenson and I.D. Wilson, Plenum Press, New York, 1994

ants may be contaminated by 1,4-dioxane [5,7], a carcinogen in rats and mice [8,9] which is absorbed through the intact skin of animals [10]. Furthermore, 1,4-dioxane has been classified as a possible carcinogen in humans [11,12]. According to the EEC directive on cosmetics [13], 1,4-dioxane must not be present in their formulations. Consequently the assay of this substance in the surfactants, used as raw materials for the production of cosmetics, and in the finished product is of direct concern to consumers.

Classical methods for the quantitative determination of 1,4-dioxane in cosmetic matrices are based on gas chromatography (GC) [4,14,15] or headspace GC [6,16,17]. These techniques, however, have distinct disadvantages such as complex and time consuming sample pre-treatment [4,14] extensive calibrations [6,16,17], the need for prolonged equilibrium times [6], unsatisfactory reproducibility [14,15] and a high degree of variability in the recovery values obtained from different cosmetic formulations [14]. For routine determination of 1,4-dioxane in cosmetics, an accurate, precise and simple method was required. This study reports on the development of solid-phase extraction procedures for the rapid and efficient purification of the complex cosmetic matrices before assay of 1,4-dioxane by reversed-phase high-performance liquid chromatography (RP-HPLC) and GC-mass spectrometry (GC-MS). The application of these methods to the determination of 1,4-dioxane in commercial surfactants and finished cosmetic products is also reported.

EXPERIMENTAL

Materials

HPLC-grade 1,4-dioxane, hexane, dichloromethane, acetonitrile and water were supplied by Farmitalia Carlo Erba (Milan, Italy). Bakerbond C-18, Bakerbond Si and Bakerbond CN cartridges were obtained from J T Baker (Phillipsburg, NJ, USA). Surfactant samples were from commercial suppliers. The nomenclature of the Cosmetic, Toiletry and Fragrance Association (CTFA) Cosmetic Ingredient Dictionary [18] has been used throughout. Commercial cosmetics, containing polyoxyethylated surfactants, were from retail stores or from manufacturers or importers of these products.

Chromatography

The HPLC apparatus consisted of a Jasco chromatographic system (Model 880-PU pump, Model 880-02 ternary gradient unit and Model 875 UV/VIS detector; Jasco, Tokyo, Japan) linked to an injection valve with a 20µl sample loop (Rheodyne, Cotati, CA, USA) and a chromatographic data processor (Chromatopac C-R3A, Shimadzu, Kyoto, Japan). The detector was set at 200 nm and 0.01 absorbance units full scale. Sample injections were made with a Hamilton 802 RN syringe (10-µl; Hamilton, Bonaduz, Switzerland).

Separations were performed on a LiChrospher CH-8 column (5-µm, 250 x 4.0 mm i.d.; Merck, Darmstadt, FRG) under gradient conditions at a flow-rate of 1.0 ml/min. Solvent A was 5% (v/v) acetonitrile in water and solvent B was 50% (v/v) acetonitrile in water. The elution programme was as follows: isocratic elution with 5% solvent B, 95% solvent A for 5 min, then a 2-min linear gradient to 95% solvent B; the mobile phase composition was finally maintained at 95% solvent B for 1 min. Samples were injected 0.5 min after the start of the elution programme. The mobile phase was filtered through type HVLP filters (0.45-µm; Millipore S.A., Molsheim, France) and on-line degassed by a model ERC-3311 automatic solvent degasser (Erma, Tokyo, Japan). Chromatography was carried out at ambient temperature. Peak areas were quantified using the integrator which was calibrated with standard solutions of pure 1,4-dioxane.

GC analyses were performed according to Black *et al* [14] using a Fractovap 4200 gas chromatograph (Carlo Erba) fitted with a flame ionisation detector. The glass column (2 m x 4 mm i.d.) was packed with Chromosorb 106 (Alltech, Eke, Belgium). The operating conditions were: column temperature, 210°; injector port temperature, 230°; detector temperature, 230°; carrier gas (nitrogen) flow-rate, 40 ml/min.

GC-mass spectrometry was carried out at an ionisation potential of 70 eV using an HP 5890 gas chromatograph (Hewlett-Packard, Avondale, PA, USA) equipped with a 5970 Mass Selective Detector (Hewlett-Packard) with the transfer line held at 280°. For quantification the instrument was operated in the selective ion-detection mode scanning masses 88 (molecular ion of 1,4-dioxane) and 92 (molecular ion of toluene, internal standard). Samples (1 μl) were introduced using cool on-column injection. A de-activated fused silica pre-column (5 m x 0.32 mm i.d.; Hewlett-Packard) and an analytical column (Poraplot Q, 25 m x 0.32 mm i.d.; Chrompack, Middelburg, The Netherlands) were used. The operating conditions were: initial temperature, 40°; ramp 40-220°, rate 30°/min; carrier gas, helium; inlet pressure, 100 Kpa.

The identity of the 1,4-dioxane peak was assigned by co-chromatography with the authentic substance. For some of the samples the identification was established by GC-MS analysis, as outlined above.

Sample Processing

A 1-g amount of the surfactant product was accurately weighed into a 20 ml volumetric flask. Water was added, the sample mixed and then diluted to the mark. A 2.5 ml aliquot of this solution was applied to a C18 cartridge. The cartridge had been pre-conditioned with 5 ml of acetonitrile and then 5 ml of water. The sample was eluted with 2 ml of acetonitrile-water (10:90, v/v). The latter fraction was directly analysed by RP-HPLC.

The cosmetic product (0.5 to 0.6 g) was accurately weighed into a 10 ml glass centrifuge tube; 4 ml of 10% (v/v) dichloromethane in hexane were added and the sample was mixed vigorously on a vortex mixer and centrifuged at 2000 g for 2 mins. The extraction was repeated with 2 ml of 10% dichloromethane in hexane and the combined supernatant solutions were applied to a pre-conditioned 500 mg CN cartridge (3 ml of acetonitrile followed by 5 ml of 10% dichloromethane in hexane) at a flow rate of ca. 1.5 ml/min. The column was then aspirated to dryness by centrifugation at 1000 g for 1 min and eluted with two 0.8 ml aliquots of 15% v/v acetonitrile in water. The eluate was passed directly through a C18 cartridge (sorbent weight, 200 mg), which had previously been primed with 3 ml of acetonitrile and 5 ml of 15% acetonitrile in water. A further 0.4 ml portion of 15% acetonitrile in water was passed through the column to complete the elution of 1,4-dioxane. The eluate from the C18 cartridge was analysed directly by RP-HPLC.

RESULT AND DISCUSSION

Surfactants

The sulphated polyoxyethylene alcohols were examined in this investigation because they represent the most commonly used polyethoxylated surfactants in a variety of cosmetic products [1,2] (e.g. shampoos, bath preparations, skin cleansing gels and lotion formulations). Prior to RP-HPLC analysis [19], a purification step was essential to remove matrix peaks in the chromatogram of the surfactant extract which interfere with the determination of 1,4-dioxane. Purification procedures based on solid-phase extraction techniques were evaluated. The octadecyl-bonded silica sorbent was found to retain 1,4-dioxane in water; removal of adsorbed 1,4-dioxane (recovery rates>96%) was obtained by elution with

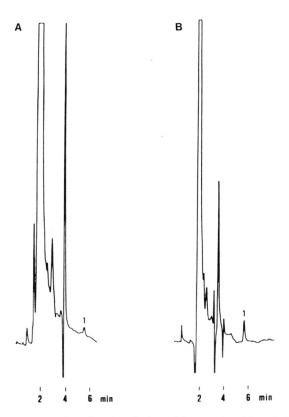

Figure 1. RP-HPLC chromatograms of a sodium lauryl ether sulphate product (A) and of a sample containing an undefined blend of fatty alcohol ether sulphates (B). Operating conditions as described in the text. Peak: 1 = 1,4-dioxane. (Reprinted with permission from J. Pharm. Biomed. Anal. [19]. Copyright 1990, Pergamon Press plc).

10% v/v acetonitrile in water. Accordingly, sample preparation consisted simply of dissolving the surfactant product in water, applying an aliquot of the sample to a C18 cartridge and eluting it with 2 ml of 10% acetonitrile in water. Typical RP-HPLC chromatograms are shown in Figure 1 of a sodium lauryl ether sulphate product (Fig.1A) and of an undefined blend of fatty alcohol ether sulphates (Fig.1B) purified according to the procedure described above and found to contain 35.6 and 113.1 µg/g of 1,4-dioxane respectively. The proposed method is more rapid (taking less than 15 min to perform) and simpler than others reported in the literature [4,15,20]; moreover multiple samples can be processed simultaneously with specially designed vacuum manifolds.

The average recovery of 1,4-dioxane (±SD) from surfactant samples was 95.7 ± 3.3% (n=10) in the 40 to 120 µg/g range. The recovery was found to be reproducible between different batches of C18 cartridges. The relative standard deviation (RSD) values for the intra-assay and inter-assay reproducibility were 3.5% (n=10) and 6.8% (n=10) respectively. The present method was compared with headspace GC [20] on the same surfactant sample. Virtually identical values were obtained, thereby proving the validity of the solid-phase extraction procedure for the assay of 1,4-dioxane in ethoxylated alcohol sulphate products.

Cosmetics

Our objective was to evaluate the utility of solid-phase extraction procedures for the rapid and simple purification of cosmetics prior to assay of 1,4-dioxane by RP-HPLC [7].

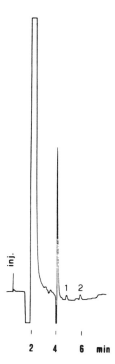

Figure 2.   RP-HPLC chromatogram of a cleansing lotion formulation. Conditions as in Fig.1. Peaks 1 = unknown compound, 2 = 1,4-dioxane. (Reprinted with permission from Analyst [7]. Copyright 1990, The Royal Society of Chemistry).

Mixtures of hexane and dichloromethane were tested as extraction solvents in combination with cartridges prepacked with polar sorbents such as silica and cyanopropyl-bonded silica. The dichloromethane content of the hexane-dichloromethane mixture strongly influenced the recovery of dioxane. Lower recoveries (<60%) were observed at dichloromethane concentrations higher than 10% v/v, owing to incomplete adsorption of 1,4-dioxane on the extraction column during sample application. Therefore dichloromethane-hexane (10:90, v/v) was used as the solvent for sample extraction. Both Si and CN cartridges were found to retain 1,4-dioxane efficiently. The CN column, however, afforded the more effective sample clean-up. The CN packing is of intermediate polarity and can be used in both normal and reversed-phase modes [21]. By this means, improved selectivity was obtained which combined the two separation principles. The eluate from the CN column was passed directly through a C18 cartridge to obtain a clear solution, suitable for injection onto the HPLC column.

Commercial cosmetic products containing no detectable dioxane were spiked at levels corresponding to 30 and 90 µg/g. The mean recoveries (n=6) for a day cream, a moisturising lotion and a shampoo were 87.0% with a RSD of 2.5%, 87.9% with a RSD of 3.1% and 86.5% with a RSD of 4.1%, respectively. In contrast, a previous investigation carried out by GC [14] produced a mean recovery of 63% accompanied by a high degree of variability (RSD, 19.9%) in the values obtained from different cosmetic formulations. The recovery of the proposed method was found to be reproducible between different batches of CN and C18 cartridges. Moreover the extraction and solid-phase isolation was achieved with small volumes of the appropriate solvent mixtures, in contrast to the purification procedure used in an earlier investigation [14].

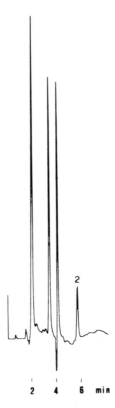

2 4 6 min

Figure 3. Chromatogram of a hair lotion preparation. Conditions and peak identification as in Figure 2.

The method developed in this study is simpler and more rapid (taking less than 25 min to perform) than the previously adopted GC [14] (analysis time, ca. 100 min) or headspace GC [6] (analysis time, ca. 16 hours) techniques as time-consuming steps including solvent partitions [14], heating [6,14], several column purifications [14] and extensive calibrations [6] are not required. Moreover multiple samples can be processed simultaneously with specially designed vacuum manifolds. Representative chromatograms of a cleansing lotion product containing 7.3 µg/g of 1,4-dioxane and of a hair lotion preparation containing 85.9 µg/g of 1,4-dioxane are shown in Figures 2 and 3, respectively. The minimum quanti-

Table 1.   1,4-dioxane content of cosmetic products determined by
RP-HPLC.

1,4-dioxane (µg/g)	Percentage of total investigated products (n=22)
n.d.	54
7-10	14
10-50	27
50-100	5

n.d. = not detected.

fiable amount was 7.0 µg/g. Applying the foregoing procedure to a cleansing lotion, 1,4-dioxane (7.3 µg/g) was determined with a RSD of 4.4% (n=10) for intra-assay reproducibility and 9.6% (n=10) for inter-assay reproducibility.

A variety of commercially available cosmetics were assayed for 1,4-dioxane according to the procedure described here. The products (n=22) included shampoos, liquid soaps, day creams, suncreams, cleansing lotions, bath foams, after shave emulsions and hair lotions. As reported in Table 1, 46% of the products investigated contained 1,4-dioxane, with levels higher than 10 µg/g in 32% of the samples.

In the latter cosmetic preparations the presence of 1,4-dioxane was confirmed by GC-MS. The solid-phase extraction procedure used for the RP-HPLC assay was found to be suitable for GC-MS analysis provided that the CN cartridge was eluted with acetonitrile instead of 15% acetonitrile in water. In fact although a high proportion of water is required in the sample injected onto the HPLC column to maintain chromatographic efficiency, the presence of water interferes with the GC analysis. The levels of 1,4-dioxane in four different cosmetic products determined by HPLC-UV and GC-MS using selected ion monitoring are reported in Table 2. The two methods produced consistent results confirming the validity of the purification procedure used. The higher values obtained by GC-MS were traced to improved recovery during sample extraction. The MS is more sensitive (minimum quantifiable amount, 2 µg/g) than the UV detector; consequently a lower amount of the cosmetic product (0.2 g) is required for the GC-MS assay and a more efficient extraction into the dichloromethane-hexane (10:90 v/v) solvent is thus achieved.

Table 2. Comparison of amounts of 1,4-dioxane in cosmetic products determined by RP-HPLC and GC-MS.

Sample	Concentration* (µg/g)	
	RP-HPLC	GC-MS
Day cream	n.d.	n.d.
Shampoo	18.4	22.3
Hair lotion	85.9	95.2
Bath foam	35.7	40.3

* = Mean value of three determinations
n.d. = not detected

The results presented in Table 1 indicate that the control of 1,4-dioxane contamination in commercial cosmetic products should be considered by national and international authorities. Moreover, it should be stressed that even when the 1,4-dioxane content is low, under normal conditions of cosmetic use, long-term applications to skin are usual.

CONCLUSIONS

Purification procedures based on the use of disposable C18 and CN cartridges have been developed for the assay of 1,4-dioxane in cosmetic matrices. Because of the minimal sample preparation, good accuracy and reproducibility the proposed methods are suitable for routine quality control analyses of polyethoxylated raw materials and finished cosmetic products.

REFERENCES

1. G. Baker. **In:** "Surfactants in Cosmetics". M.M. Rieger (ed.), Dekker, New York, pp251-292 (1985).
2. P. Thau. **In:** "Surfactants in Cosmetics". M.M. Rieger (ed.), Dekker, New York, pp349-376 (1985).
3. D.J. Worsfold and A.M. Eastham. Cationic polymerisation of ethylene oxide. *J. Am. Chem. Soc.* 79, 897-899 (1957).
4. T.J. Birkel, C.R. Warner and T. Fazio. Gas chromatographic determination of 1,4-dioxane in polysorbate 60 and polysorbate 80. *J. Assoc. Offic. Anal. Chem.* 62, 931-936 (1979).
5. J.A. Wenninger. *Assoc. Food Drug Off. U.S.Q. Bull.* 44, 145-152 (1980).
6. S.C. Rastogi. Headspace analysis of 1,4-dioxane in products containing polyethoxylated surfactants by GC-MS. *Chromatographia* 29, 441-445 (1990).
7. S. Scalia, M. Guarneri and E. Menegatti. Determination of 1,4-dioxane in cosmetic products by high-erformance liquid chromatography. *Analyst* 115, 929-931 (1990).
8. C. Hoch-Ligeti, M.F. Argus and J.C. Arcos. Induction of carcinomas in the nasal cavity of rats by dioxane. *Br. J. Cancer* 24, 164-170 (1970).
9. National Cancer Institute, Technical Report series no.80, NCI-CG-TR-80, U.S. Dept. Health, Education and Welfare, Washington DC (1978).
10. H. Leung and D.J. Paustenbach. Cancer risk assessment for dioxane based upon a physiologically-based pharmacokinetic approach. *Toxicol. Letts.* 51, 147-162 (1990).
11. International Agency for Research on Cancer. "Monographs on the Evaluation of Carcinogenic Risk of Chemicals to Humans" vol.11, World Health Organisation (ed.), Lyon, pp247-256 (1976).
12. I. Lundberg, J. Hogberg, T. Kronevi, B. Holmberg. Three industrial solvents investigated for tumour promoting activity in the rat liver. *Cancer Letts.* 36, 29-33 (1987).
13. European Economic Community Council Directive 76/768/EEC. Appendix II (1976).
14. D.B. Black, R.C. Lawrence, E.G. Lovering and J.R. Watson. Gas-liquid chromatographic method for determining 1,4-dioxane in cosmetics. *J. Assoc. Offic. Anal. Chem.* 66, 180-183 (1987).
15. B.A. Waldam. Analysis of 1,4-dioxane in ethoxylated compounds by gas chromatography/mass spectrometry using selected ion monitoring. *J. Soc. Cosmet. Chem.* 33, 19-25 (1982).
16. H. Helms. A quick and economical method for the determination of 1,4-dioxane and ethylene oxide in cosmetic preparations. *Parf. Kosmet.* 69, 17-18 (1988).
17. H. Beernaert, M. Herpol-Borremans and F. De Cock. Determination of 1,4-dioxane in cosmetic products by headspace gas chromatography. *Belg. J. Food Chem. Biotechnol.* 42, 131-135 (1987).
18. N.F. Estrin. "Cosmetic Ingredients Dictionary" 3rd edn., The Cosmetic, Toiletry and Fragrance Association, Washington (1982).
19. S. Scalia. Reversed-phase high-performance liquid chromatographic method for the assay of 1,4-dioxane in sulphated polyoxyethylene alcohol surfactants. *J. Pharm. Biomed. Anal.* 8, 867-870 (1990).
20. M. Ronza, A. Ariotto and M. Principi. Sulphation of ethoxylated alcohol - optimisation of sulphation technology to reduce 1,4-dioxane in the final product. *La Revista delle Sostanze Grasse* 65, 497 (1988).
21. M. De Smet, G. Hoogewijs, M. Puttemans and D.L. Massart. Separation strategy of multicomponent mixtures by liquid chromatography with a single stationary phase and a limited number of mobile phase solvents. *Anal. Chem.* 56, 2662-2670 (1984).

COLUMN SWITCHING STRATEGIES FOR COMPLEX PETROCHEMICAL SAMPLE

ANALYSIS

Andrew J. Packham

Technical and Analytical Solutions
Pelham House
49 Pelham Street
Ashton Under Lyne
Lancashire
OL7 0DT. UK

SUMMARY

The development of a method for the analysis of a specific, or group of specific analytes, in a complex matrix is difficult. Problems arise due to the large number of interfering substances present and because the analyte of interest is often only present at low concentrations. Column switching techniques have been commonly used to solve this type of analytical problem. As the number of different processes whereby the separation of sample components may be achieved increases the ability to handle complex samples likewise increases. As there are a large number of different parameters that require either complete or partial optimisation a significant demand is placed on the method developer.

The environmental impact of petrochemicals is accepted, their role in the initation of cancer has been clearly shown. The analysis of petrochemicals, and the product formed through chemical or combustive modification, for the specific carcinogenic substances is thus of importance. Procedures whereby the polynuclear aromatic hydrocarbons in petrochemical samples, of varying complexity, may be analysed are discussed. Each sample, having different properties, gives different problems which require different column configurations.

INTRODUCTION

High performance liquid chromatography, (HPLC), is often the method of choice in many analytical laboratories. However, due to the large number of components in a complex sample, these samples may be difficult to analyse by orthodox liquid chromatographic techniques. The problems associated with analysing complex samples have been shown clearly by Davis and Giddings [1] Cortes [2] and Guiochon *et al* [3] who estimated that a column efficiency of around 200,000 theoretical plates was required to give a 90% probability that a

Sample Preparation for Biomedical and Environmental Analysis,
Edited by D. Stevenson and I.D. Wilson, Plenum Press, New York, 1994

227

20 component mixture would be fully resolved. When it is considered that a complex sample may contain hundreds of components the analytical difficulties can be clearly seen.

HPLC has been used widely to analyse an extensive range of mixtures for a variety of analytes. For a simple mixture, containing not more than 10 to 20 components of similar chemical nature, the development of a method is relatively straightforward. As the number of components increases, and the components become chemically more diverse, the effort required multiplies.

A complex sample can be defined as a mixture in which one or more of the constituents hampers the analysis of the substance of interest. It is possible to sub-divide complex samples into two broad categories; numerically and chemically complex mixtures. Numerically complex samples are complex by virtue of the large number of components present while chemical complexity is due to the presence of components that are incompatible with the analytical method used.

Due to new legislation, by both national and international bodies, increased awareness of the environmental impact of many substances and an ever growing consumer pressure for more 'healthy and natural' products the analyst is required to examine a wider range of complex mixtures for ever decreasing concentrations. Complex samples are found in every area of analytical science. Biomedical samples may contain protein which is incompatible with most HPLC columns, pharmaceutical preparations such as creams and ointments are not in a state suitable for analysis. Environmental samples often contain very low levels and therefore require extensive concentration. Crude petroleum contains a vast range of components while formulated oils contain not only a large number of components but also additives which hamper their analysis by HPLC. Water contaminated with oil is probably one of the most difficult matrices to analyse, not only is it both chemically and numerically complex but it is also very difficult to sample reproducibly.

By definition, the analysis of a complex mixture is complicated and implies that many separate stages must be followed to obtain useful information. Each stage introduces error into the procedure, that in total, may account for a significant percentage of the overall error. In extreme cases, the sample which is finally analysed may bear no relation to the original if it has passed through a number of pre-treatment stages. These pre-treatment stages are required to either simplify the sample matrix or to increase the concentration of the analyte to a level where the selected analytical procedure may reliably determine the concentration. Methods such as liquid/liquid or solid phase extraction, preparative thin layer chromatography (TLC), column chromatography, dialysis and filtration can be realistically incorporated within one combined system by using column switching techniques.

Column switching involves the coupling of two or more columns together by switching valves, selecting the appropriate valve, and actuating it at the correct time. This causes different fractions of the sample to follow different paths through the column network. The separation effort can then be concentrated on the important fractions whilst ignoring those fractions that are of no interest to the analyst. The technique usually relies on an initial fractionation, or incomplete resolution, of the sample on one column. Subsequently the fraction containing the substances of interest is passed, for resolution, to a second column.

The separation of the sample consequently occurs, not just by chromatographic means, but also by physical means, which are not related to chemical properties of the analyte and are under the control of the analyst, these dual modes of separation may result in very efficient separations. Column switching techniques may be used to analyse a sample directly, without any pre-treatment stages, and so reduce the chance of laboratory or personal contamination as the entire separation may be performed within an enclosed environment. It is also much easier to automate a separation method relying on column switching. Methods using a number of chromatographic separations to clean-up the sample can often be converted to a column switching configuration with the minimum of difficulty. The resulting increase in safety and the improved analytical speed, precision and accuracy often make up for the initial cost of the additional equipment required.

As with all methods there are always limitations. The major problem that will be encountered is likely to be the incompatibilities that occur between different solvents. Other limitations to the method are related to its complexity and the need for accurate control. The switching valves, automated actuators and all the associated controlling equipment are costly. The requirements for accurately timed switching means that a computer based system is not a luxury but a necessary requirement of the system. Another consequence of complexity is that method development and subsequent optimisation is difficult and time consuming.

The use of column switching started in 1973 with the development of a low dead volume switching valve that would maintain the efficiency of the system [4] but until the early 1980s the rate of publications remained low. However, reflecting the increased general awareness of the need to analyse complex mixtures, the number of publications has risen rapidly in the mid and latter parts of the '80s. The upward trend is continuing into the 1990s for a number of reasons. New columns are becoming available and many of these will have an impact on multi-mode separations with the same mobile phase. In addition the use of computer based equipment allows the simpler and more accurate control of complex column networks. Indeed the price of many personal computer based chromatographic data stations now rivals that of less flexible chromatographic integrators. New legislation will still remain the driving force behind the need to analyse complex mixtures for environmentally damaging or harmful substances.

Separation Of Polynuclear Aromatic Hydrocarbons

Polynuclear aromatic hydrocarbons (PNAs) are a group of common environmental pollutants thought to cause a number of different types of cancers. The measurement of this group is thus of extreme importance and spans not only the analysis of base petroleum products such as aviation, heating and diesel oils but also environmental samples such as water, sediment and air extracts. Each of the different matrices presents its own special problems to the analyst. For example, the PNAs are present in diesel oil at significant levels eliminating the need for a pre-concentration stage but, as the sample is very complex, requiring a number of fractionation steps. Drinking water, on the other hand, usually contains low concentrations of PNAs, and thus pre-concentration rather than fractionation is required.

The optimum column switching configuration usually involves the selection of columns with very different separation mechanisms. This ensures that the components of the mixture that co-elute with the analyte during the fractionation, and therefore are transferred to the next column during zone transfer, are separated from the analyte on that column. The greater the difference in retention mechanism, the more likely it is that a column switched configuration will be successful. However, many chromatographic systems require solvents that are incompatible with those required for other modes of separation, for example normal and reversed-phase, ion exchange and organic size exclusion.

One of the few attempts to directly couple normal and reversed-phase separations was reported by Sonnefeld et al [5]. These workers used a multi-column separation technique for the determination of PNAs in shale oil. The incompatibilities occurring between the normal and reversed-phase solvents were overcome by using a diamine column between the incompatible systems. The PNAs were concentrated on this column after fractionation on an amino-silane column, the normal phase solvent was then removed by a stream of dry nitrogen. The PNAs were then eluted from this column and separated using gradient elution in a reversed-phase system. No significant difference was found between this method and an off-line technique.

The on-line connection of size exclusion to reversed-phase has also been achieved

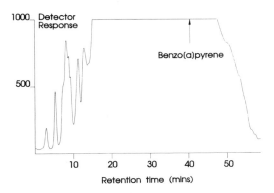

Figure 1. Attempted separation of 10 μl of aviation fuel in a single C18 (C4) column (Vydac TP102 25cm x 4.6mm). Analytical conditions : linear gradient over 45 minutes, 30% to 100% acetonitrile, 1 ml/min detection by fluorescence ex 254nm, em 450nm.

[6,7]. In these methods the aqueous miscible organic mobile phase was mixed on-line with water to reduce the organic content and permit reconcentration at the head on the next column.

The use of chiral to reversed-phase systems has been used to remove the interferences from the analytes [8]. The mobile phase must be carefully selected so that not only are the majority of the interferences retained on the fractionation column (cyclodextrin) but also the amount of organic modifier in the transferred zone must not prevent re-concentration on the next column. The complex nature of the retention mechanism of the cyclodextrin bonded phases [9] produces significant separation between the PNAs and the majority of the interferences.

Aviation Fuel

The need for sample pre-treatment for aviation fuel is clearly illustrated in Figure 1. The complexity of aviation fuel, which is a relatively simple petroleum product, is still too great for a direct analysis by HPLC. In contrast Figure 2 shows the separation of the sample with a simple two column procedure. The aviation fuel was directly injected into the chromatograph and the sample fractionated on a τ cyclodextrin column. The portion of the

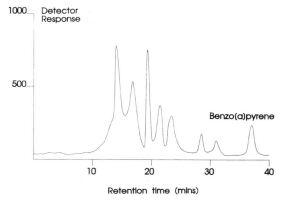

Figure 2. Determination of benzo[a]pyrene in aviation fuel. (Sample spiked with 8 ng per injection). Frac – tionation conditions : τ cyclodextrin (C2), 40% acetonitrile in water, 1 ml/min^{-1}. 45 second heart cut starting at 3 minutes 45 seconds. Analytical conditions : C18 (C4) column (Vydac TP102 25cm x 4.6mm), linear gradient over 45 minutes, 30% to 100% acetonitrile, 1 ml min detection by fluorescence ex 254nm, em 450nm.

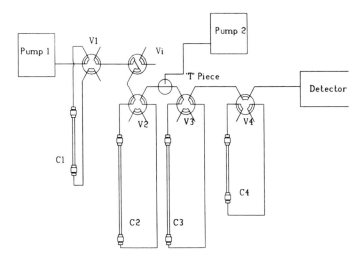

Figure 3.   Column configuration used for column switched HPLC, for details see text.

eluent which contained benzo[a]pyrene, a potent pro-carcinogen, was transferred through a switching valve to a C18 column for separation. The column configuration used to obtain the results discussed here is shown in Figure 3. The system consisted of two pumps, a Waters 6000 and a Kontron 735, 4 switching valves (Rheodyne 7001), 4 columns, and an injection valve (Rheodyne 7020) and a 'T' piece. Detection was by both fluorescence and ultra-violet. Fluorescence (Perkin Elmer Model 3000) was used to obtain the selectivity and sensitivity while UV (Waters Lambda Max) was used as a confidence check. Valve 1 and column 1 (5 cm x 4.6 mm, Spherisorb C18, 5 µm) were used when pre-concentration was required. When this was not required an amino column (5 cm x 4.6 mm, Hypersil NH_2) was used which showed minimal retention and acted as a guard column for the system. Columns C2 and C3, which were selected by valves V2 and V3, provided different mechanisms for samples fractionation. C2 was a β cyclodextrin (25 cm x 4.6 mm) and C3 was a τ cyclodextrin (25 cm x 4.6 mm). Samples could be fractionated on either of these columns. When very complex samples were used fractionation on both columns was found to be necessary. To ensure that the heart cut was well retained of the head on the second column water or buffer was introduced through the 'T' piece. This reduced the organic modifier percentage and so increased retention. It was necessary to ensure the head reconcentration occurred to maintain the efficiency of the whole chromatographic system. Column 4, controlled by V4, was used to separate the samples. Normally a 15 cm x 4.6 mm Vydac C18 TP201 column was used but occasionally the length was increased to 25cm. The Vydac column used is routinely used to separate the 16 PNA EPA 610 mixture.

It is interesting to note that while cyclodextrin columns have been used successfully to separate PNAs [10,11], the conditions used here, i.e. relatively high organic modifier percentages, would not result in a separation of the group. This is highly advantageous as it simplifies the operation of the system. As the PNAs elute in one, unresolved, peak only one heart cut operation is required to transfer the group onto the second analytical column where they are separated.

Lubricating Base Oils

It is also possible to use a two column configuration to determine a range of analytes in petroleum products. The determination of a range of PNAs in a lubricating base oil that

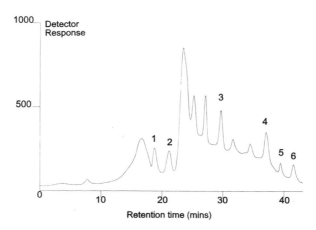

Figure 4.   Determination of PNAs base lubricating oil (30% aromatic). Fractionation conditions : β cyclodex trin (C3), 40% acetonitrile in water, 1 ml/min. 45 second heart cut starting at 3 minutes 45 seconds. Analytical conditions : C18 (C4) column (Vydac TP102 25cm x 4.6mm), linear gradient over 45 minutes, 30% to 100% acetonitrile, 1 ml/min detection by fluorescence ex 254nm, em 450nm.

had contained about 30% aromatic substances is shown in Figure 4. Any further increase in complexity requires a second fractionation stage to be incorporated. The chromatogram obtained using a three column system to analyse a 40% aromatic base oil can be seen in figure 5. To maintain the efficiency of the separation it was necessary to ensure that at, each fractionation, the fraction of interest was re-concentrated at the head of the next column. This was usually achieved by equilibrating the second column with a solvent of lower eluotropic strength than that carrying the fraction onto the column. This was, however, difficult to achieve when using the cyclodextrin columns due to the low density of the bonded phase. The amount of organic modifier necessary to elute the fraction suppressed any re-concentration that may have occurred. Without re-concentration the PNAs pass onto the second column and begin to undergo separation. As the fraction transferred is not transferred as a tight band, the fraction passed to the analytical column will need to be large to include all the PNAs. This will therefore also include a large proportion of interfering material. In addition, the time required for the heart cut to be loaded onto the analytical column was such that re-concentration did not occur on this column either. To ensure that the necessary re-concentration did occur water was mixed, on line, before the second fractionation stage to reduce the percentage organic modifier in the stream and so promote the required effect. Further experimental details are presented elsewhere [12].

Exhaust Emissions

Particulate pollution is common in all industrial areas and is a cause for concern not only due to the odour and soiling effect but also due to toxic consequences of particulate inhalation. Exhaust particles may have a diameter of less than 1 micrometer, allowing the particles to be deposited deep within the lung, predominantly within the alveolar region. The young, elderly and those with asthmatic, allergenic or cardiovascular diseases are particularly at risk to the effects on inhalation.

The complexity and nature of particulate samples makes analysis difficult. The PNAs must not only be removed from the particles but also must be separated from the interfering substances present. Current methods rely on a Soxhlet extraction, often followed by a sample clean-up stage such as solid or liquid extraction. This is necessary to remove the interferences and to produce a final sample in which the PNAs have been suitably concentrated.

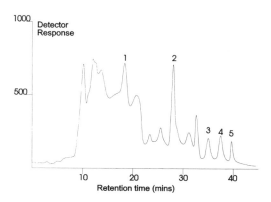

Figure 5. Determination of PNAs in base lubricating oil (40% aromatic) using a three column system. Frac tionation conditions : τ cyclodextrin (C2) and @ cyclodextrin (C3), 40% acetonitrile in water, 1 ml/min. 45 second heart cut starting at 3 minutes 45 seconds and 2 minute heart cut starting a five minutes. Analytical conditions : C18 (C4) column (Vydac TP102 25cm x 4.6mm), linear gradient over 45 minutes, 30% to 100% acetonitrile, 1 ml/min detection by fluorescence ex 254nm, em 450nm.

These methods are often complex and are likely to introduce errors due to sample loss and degradation. I.L Davies *et al* [13] used an 8 hr Soxhlet extraction with benzene prior to analysis by a coupled microbore HPLC-capillary GC system. The normal phase HPLC system was used to separate the aromatic fraction from the non-aromatic components, the aromatic fraction was then backflushed onto the GC column by means of a 25-m retention gap. Obuchi *et al* [14] describe a method to study particulate PNAs by HPLC. Following an 18 hr Soxhlet extraction into dichloromethane and fractionation on a C18 solid phase extraction column (Sep-Pak) detection limits in the picogram range were achieved. Solid phase cartridges (Bond-Elut) were also used by Karlesky *et al* [15] to determine PNAs in refinery air samples. After a 24 hr extraction into cyclohexane, the sample was extracted with N,N-dimethylformamide-H_2O and then back extracted into cyclohexane. Final analysis was by capillary GC. Airborne PNAs were measured by Choudhury and Bush [16]. Soxhlet extraction into benzene, followed by liquid-liquid extraction and TLC fractionation were employed to obtain a fraction suitable for HPLC and GC/MS analysis. All the methods are characterised by the use of a number of steps, often using toxic solvents and an extended pretreatment time.

To improve the speed of the analytical procedure, and to reduce the amount of toxic solvents required, a column switching configuration was developed. Pre-concentration of the sample allowed the preferential retention of the analytes from a large volume of sample. The chromatogram obtained from the particulate extract of the exhaust emitted by a small 3 door car with a well tuned engine (1 l engine size) fuelled by a commercially available standard un-leaded petrol is shown in Figure 6. Peak identification was by comparison of retention times and confirmed by standard addition. Quantification of the PNAs present was by comparison of peak heights with a calibration line constructed from the PNA calibration mixture (Table 1). The sampling system used in this study was rudimentary and no attempt was made to control emission flow rate or to measure other emission parameters. The emission from an exhaust is complex and contains water vapour, gaseous components, and particulate matter of various sizes. The sample used in this study is probably only a broad approximation of the true profile of emissions. Ideally a proper engine testing facility should be employed where the operational parameters can be controlled more closely. The use of a more sophisticated sampling system should allow the accurate quantification of PNAs from particulates.

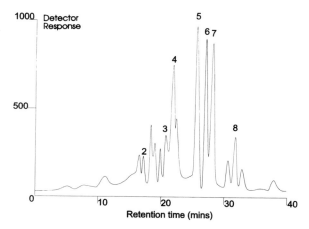

Figure 6.   Determination of PNAs from an extract of particulate emission from a 1 1 standard petrol engine running un-leaded fuel. 2 minute emission collect and glass wool pad. Extracted with 50 ml hexane for 5 minutes and evaporated, redissolved in acetonitrile (500 μl), acetonitrile evaporated and water added to make 50 μl acetonitrile, 450 μl water. Entire sample pre-concentrated on hypersil C18 cartridge (C1) and then eluted onto C18 (C4) column (Vydac TP102 25cm x 4.6mm), linear gradient over 45 minutes, 30% to 100% acetonitrile, 1 ml/min detection by fluorescence ex 254nm, em 450nm.

The automation of this process through the ASTED system provided a method whereby the PNA content of particulates may be performed on a routine basis by technical staff in safety, and with the minimum of effort.

Table 1.   Quantitative analysis of PNAs obtained from a 1 litre petrol engine.

	Polynuclear aromatic hydrocarbon	Amount present (ng)
1.	Fluoranthene	Not detected.
2.	Pyrene	20.6
3.	Chrysene	6.4
4.	Benzo(a)anthracene	4.0
5.	Benzo(a)fluoranthene	4.9
6.	Benzo(a)fluoranthene	6.5
7.	Benzo(a)pyrene	2.6
8.	Benzo(g,h,i)perylene	3.4

CONCLUSION

Column switching methodologies when applied to HPLC can achieve many benefits. They not only allow the automation or simplification of existing methods but also allow the development of otherwise demanding analyses. Column switched HPLC does however suffer from a number of problems, the thrust of further work must not only be in development of methods *ad hoc*, but must be in the formation of a generalised theory by which methods may be developed and subsequently optimised in a more coherent fashion.

ACKNOWLEDGMENTS

Technical & Analytical Solutions would like to thank the following companies and organisations for their help and assistance, Technicol Ltd, Stockport; Comus Instruments, Kingston Upon Hull, Fisons Instruments (VG Elemental), Winsford and Anachem Ltd, Luton.

Some of the work discussed here was undertaken during my Ph.D under the supervision of Dr. Peter Fielden at the Department of Instrumentation and Analytical Science, UMIST which was funded by the Science and Engineering Research Council and ESSO Petroleum Company, Abingdon.

I am also indebted to the Chromatographic Society.

REFERENCES

1. J.M. Davis and J.C. Gidding. Statistical-theory of component overlap in multicomponent chromatograms. *Anal. Chem.* 55, 418-424 (1983).
2. J.C. Giddings. **In:** Multidimensional Chromatography. Techniques and Applications, (H.J. Cortes, ed.), Marcel Dekker Inc, New York, 1990.
3. G. Guiochon, M.F. Gonnord, M. Zakaria, L.A. Beaver and A.M. Aiouff. Theoretical investigation of the potentialities of the use of multidimensional columns in chromatography. *J. Chromatogr.* 225, 415-437 (1983).
4. J.F.K. Huber, R. Van der Linder, E. Ecker and M. Oreans. Column switching in HPLC. *J. Chromatogr.,* 83, 267-277 (1973).
5. W.J. Sonnefeld, W.H. Zoller, W.E. May and S.A. Wise. On-line multidimensional liquid chromatographic determination of polynuclear aromatic hydrocarbons in complex samples. *Anal. Chem.* 54, 723-727 (1982).
6. F. Erni, H.P. Keller, C. Morin and M. Schmitt. Application of column switching in HPLC to on-line sample preparations for complex separations. *J. Chromatogr.* 204, 65-76 (1981).
7. R.A. Williams, R. Macrae and M.J. Shepherd. Non-aqueous size exclusion chromatography coupled on-line to reverse phase HPLC. *J. Chromatogr.* 477, 315-325 (1989).
8. P.R. Fielden and A.J. Packham. Selective determination of benzo(a)pyrene in petroleum-based products using multi-column liquid chromatography. *J. Chromatogr.* 497, 117-124 (1989).
9. P.R. Fielden and A.J. Packham. Retention of benzo(a)pyrene on cyclodextrin boned phases. *J. Chromatogr.* 516, 355-364 (1990).
10. D.W. Armstrong, D. Wade, A. Alak, W.L Hinze, T.E. Riehl and K.H. Bui. Liquid chromatographic separation of diasereomers and structal isomers on cyclodextrin bonded phases. *Anal. Chem.,* 57, 234-237 (1985).
11. M. Olsson, L.C Sander and S.A. Wise. Comparison of LC selectivity for polycyclic aromatic hydrocarbons on cyclodextrin and C-18 bonded phases. *J. Chromatogr.* 477, 277-290 (1989).
12. A.J. Packham and P.R. Fielden. Column switching for the HPLC analysis of polynuclear aromatic hydrocarbons in petroleum products. *J. Chromatogr.,* 552, 575-582 (1991).
13. I.L Davies, M.W. Raynor, P.T. Williams, G.E. Andrew and K.D. Bartle. Application of automated on-line microbore HPLC capillary GC to diesel exhaust particulates. *Anal. Chem.* 59, 2579-2583 (1987).
14. A. Obuchi, H. Aoyama, A. Ohi and H. Ohuchi. Determination of polycyclic aromatic hydrocarbons in diesel oil particulate matter and diesel fuel-oil. *J. Chromatogr.* 312, 274-283 (1984).
15. D.L. Karlesky, M.E. Rollie, I.M. Warner and C. Ho. Sample clean-up for polynuclear aromatic compounds in complex matrices. *Anal. Chem.* 58, 1187-1192 (1986).
16. D.R. Choudhury and B. Bush. Chromatographic-spectrometric identification of airborne polynuclear aromatic hydrocarbons. *Anal. Chem.* 53, 1351-1356 (1981).

AUTHOR INDEX

COMPOUND INDEX

phorate sulfone, 197
phosphoethanolamine, 106
o-phthalaldehyde, *see* orthophthalaldehyde
o-phthaldialdehyde, *see* orthophthaldialdehyde
orthopolychlorinated biphenyl (PCB), 191–201
polymethylacrylate, 5
polyoxyethylene alcohol,sulfated, 221
polystyrene-divinylbenzene, 155
prochloraz, 176
proline, 100, 102, 106
promazine, 164
pronase, 76
propan-2-ol, 149, 150
propionic acid, 94
propranolol, 54–58, 164
   structure, 54
protease, 76
purine, 24
pyrene, 8, 9, 234
pyridine hydrochloride, 176

ranitidine, 168

salbutamol, 164
santonin, 16
serine, 101, 105, 106
simazin, 197
sotalol, 164
styrene-divinylbenzene, 4
sublimaze, *see* fentanyl
succinate, 39
sulfatase, 173
sulfhydryl amino acid, 88, 89
sulfonamide, 7
sulfonic acid, 7
sulfosalicylic acid, 97, 103

tansy, 15, 16
taurine, 106

terbutaline, 164

tetrachlorophenol, 3
tetrahydrofuran, 80, 168
tetraphenylboron, 88
tetrazine, 174
thidiazuron, 178
   structure, 179
thioproline, 102
thiram, 30, 31
threonine, 109
thymine, 5
thimolol, 164
TPB, 101, 102
b-trenbolone, 5
triallate, 197, 198
triazine, 197
2,4,6-tribromoaniline, 178
2,4,6-trichlorophenol, 3, 4, 176
2,4,6-trichlorophenoxy moiety, 176
trichlorophenyloxalate, 32
trifluralin, 197, 198
trimethylamine N-oxide, 48
tryptophan, 24, 100, 105, 178, 180
trospectomycin, 81
tyrosine, 89

uracil, 5
urea, 140, 141

valproic acid, 8, 143–145
   structure, 143
vanoxerine, see GBR*12909*
vitamin K, 24, 25

warfarin, 24

zinc, 8

SUBJECT INDEX